PHYSICS RESEARCH AND TECHNOLOGY

QUANTUM FIELD THEORY AND APPLICATIONS

Physics Research and Technology

PHYSICS RESEARCH AND TECHNOLOGY

QUANTUM FIELD THEORY AND APPLICATIONS

NATALE PALERMA

nova science publishers

www.novapublishers.com

Library of Congress Cataloging-in-Publication Data

ISBN: 978-1-68507-930-7

Published by Nova Science Publishers, Inc. † New York

Contents

PREFACE

Quantum field theory (QFT) is a theoretical framework that combines classical field theory, special relativity and quantum mechanics, and represents the forefront of theoretical physics. The four chapters of this book include new research on quantum field theory and present different perspectives on the subject. Chapter One applies the notion of clothing particles in QFT to the theory of nuclear forces and quantum electrodynamics. Chapter Two is based on results from black hole thermodynamics at all energy scales and demonstrates that there is a natural ultraviolet applicable boundary distant from the Planck scales in QFT. Chapter Three presents a version of quantum field theory based on discrete energy-momentum and illustrates this theory using examples of fermion and boson fields. Lastly, Chapter Four discusses in detail whether or not perturbative expansion series in quantum electrodynamics are convergent and points out that the disagreement between the experimental value of the muon g-2 and the theoretical one predicted in the standard model might be due either to the larger hadronic vacuum polarization contributions or to an unexpectedly large sum of the perturbative expansion series.

As explained in Chapter 1, notion of clothing particles in quantum field theory (QFT), put forward by Greenberg and Schweber and developed by M. Shirokov, is applied in the theory of nuclear forces and quantum electrodynamics (QED). Along this guideline the authors have built up novel families of Hermitian, relativistic energy-independent and nonlocal interactions between: 1) the clothed mesons and nucleons (antinucleons) in case of mesodynamics (the *N-N* scattering, the quasipotential for *2N*- and *3N*- forces with relativistic description of the deuteron and triton properties and elastic p-d scattering) 2) the clothed electrons (positrons) and photons, responsible for different processes in QED (the electron-electron scattering, the Compton effect, the electron-positron annihilation, the description of the para-positronium properties).

Based on the results from black hole thermodynamics at all energy scales, Chapter 2 demonstrates that, both for the discrete QFT previously introduced by the author and for QFT in continuous space-time, there is a natural ultraviolet applicable boundary (cut-off) distant from the Planck scales. It is important that this boundary exists irrespective of the fact in which pattern, perturbative or non-perturbative mode, QFT is studied. Different inferences from the obtained results are discussed, some statements are revised.

Chapter 3 presents a version of quantum field theory based on discrete energy-momentum. Beginning with annihilation and creation operators, the authors construct interaction quantum number operators. These are employed to define discrete quantum fields. The authors then discuss scattering operators and their properties. The theory is illustrated using examples of fermion and boson fields.

Whether perturbative expansion series in quantum electrodynamics (QED) are convergent or not is discussed in detail, by taking the radiative corrections to the magnetic moment of a charged Dirac fermion such as the muon as an example. In Chapter 4, it is shown that they are asymptotic expansion, convergent, or divergent series, depending on the fine-structure constant and the total number and masses of the charged particles. It is remarkable that their convergence not only constraints the fine-structure constant and the total number and masses of the fermions, but also forbids the existence of the magnetic monopole. It is also pointed out that the disagreement between the experimental value of the muon g-2 and the theoretical one predicted in the standard model, if it stayed, might be due either to the larger hadronic vacuum polarization contributions or to an unexpectedly large sum of the perturbative expansion series.

In: Quantum Field Theory and Applications ISBN: 978-1-68507-930-7
Editor: N. Palerma

Chapter 1

THE REPRESENTATION OF CLOTHED PARTICLES IN QUANTUM FIELD THEORY

***A. Arslanaliev*[1,2], *Y. Kostylenko*[1] *and A. Shebeko*[1,*]**
[1]Institute of Theoretical Physics, NSC "Kharkov Institute of Physics & Technology", Kharkiv, Ukraine
[2]V.N. Karazin Kharkiv National University, Kharkiv, Ukraine

Abstract

Notion of clothing particles in quantum field theory (QFT), put forward by Greenberg and Schweber and developed by M. Shirokov, is applied in the theory of nuclear forces and quantum electrodynamics (QED). Along this guideline we have built up novel families of Hermitian, relativistic energy-independent and nonlocal interactions between: 1) the clothed mesons and nucleons (antinucleons) in case of mesodynamics (the $N - N$ scattering, the quasipotential for $2N$- and $3N$- forces with relativistic description of the deuteron and triton properties and elastic p-d scattering) 2) the clothed electrons (positrons) and photons, responsible for different processes in QED (the electron-electron scattering, the Compton effect, the electron-positron annihilation, the description of the para-positronium properties).

PACS: 11.10.-z, 12.20.-m, 12.38.-t, 21.30.-x, 21.45.-v

---

[*]Corresponding Author's Email: shebeko@kipt.kharkov.ua.

Keywords: unitary clothing transformations, clothed particles, quantum electrodynamics, nuclear forces, quantum chromodynamics, off-energy-shell effects, few-body systems

1. Introduction

During the last twenty years we have seen a number of successful applications of the so-called method of unitary clothing transformations (henceforth, the UCT method) in nuclear physics starting from popular models for interacting meson and nucleon fields (see survey [32] and refs therein). Being based upon the notion of clothed particles (with properties of physical ones) put forward by Greenberg and Schweber (1958) [10, 29], which survived its second birth in the 70s owing to the research work by Michail Shirokov [38] and developed then by the collaboration of the Institute for Theoretical Physics (Kharkov, Ukraine), Laboratory of Theoretical Physics (Dubna, Russia) and INFN (Univ. Padova, Italy) [36, 37, 17, 16]. For the past ten years we have extended our research work to quantum electrodynamics (QED) [4], in particular, when describing the positronium properties [18].

Recall that the UCT method itself is to express the total field Hamiltonian H and other operators of great physical meaning, e.g., the Lorentz boost generators and current density operators that depend initially on the creation and destruction operators for the bare particles, through a set of their clothed counterparts. It is achieved via special unitary clothing transformations that remain the Hamiltonian intact. In the course of the clothing procedure, a large amount of virtual processes associated in our case with the meson absorption/emission, the nucleon-antinucleon pair annihilation/production and other cloud effects turns out to be accumulated in the creation (destruction) operators for the clothed particles. Such a bootstrap reflects the most significant distinction between the concepts of clothed and bare particles.

It is proved that the UCT method is sufficiently flexible being applied not merely to local field models including ones with derivative couplings and spins $j \geq 1$. Within the approach, all interactions constructed are responsible for physical (not virtual) processes in a given system of interacting fields. Such interactions are Hermitian and energy independent (that makes them convenient for nuclear calculations) including the off-energy-shell and recoil effects (the latter in all orders of the $1/c^2$- expansion). In particular, we have managed to build up a new family of such interactions in the system of bosons (π-, η-, ρ-, ω-, δ- and σ-mesons) and fermions (nucleons and antinucleons). Besides, the interaction operators for processes of the type $2N \to 2N$, $bN \to bN$, $3N \to 3N$ and $NN \leftrightarrow bNN$ are derived on one and the same physical footing. As a whole, persistent clouds of virtual particles are no longer explicitly contained in the clothed-particle representation, and their influence is included in the properties of clothed particles (these quasiparticles of the UCT method). In addition, we would like to stress that the problem of mass and vertex renormalization is intimately interwoven with constructing interactions between clothed particles. Renormalized quantities are calculated step by step in the course of clothing procedure unlike some approaches where they are introduced inconsistently.

These common features are suggested to be employed not only in mesodynamics but in other field models, e.g., quantum electrodynamics (QED) and quantum chromodynamics (QCD). Therefore, not occasionally that the UCT method has been turned out to be useful when describing properties of the positronium, the simplest bound state in QED.

2. Some Recollections

The clothing procedure itself is realized along the chain of transformations: bare particles with bare masses $\to$ bare particles with physical masses $\to$ physical (observable) particles [16, 8]. Such approach is useful for drawing some parallels between the UCT method and the method of canonical transformations (in particular, the Bogoliubov-type ones) in the theory of superfluidity and superconductivity [24]. In this context, we start with the Lagrangians and Hamiltonians for interacting boson (scalar, pseudoscalar, vector) and fermion (electron, nucleon) fields by using the well-known Yukawa-type couplings, viz.,

$$\mathcal{L}_s = -g_s^0 \bar{\Psi}\Psi\Phi_s, \tag{1}$$

$$\mathcal{L}_{ps} = -ig_{ps}^0 \bar{\Psi}\gamma_5\Psi\Phi_{ps}, \tag{2}$$

$$\mathcal{L}_{\text{v}} = -g_{\text{v}}^0 \bar{\Psi}\gamma_\mu\Psi\Phi_{\text{v}}^\mu - \frac{f_{\text{v}}^0}{4m^0}\bar{\Psi}\sigma_{\mu\nu}\Psi\Phi_{\text{v}}^{\mu\nu}, \tag{3}$$

where $\sigma_{\mu\nu} = \frac{i}{2}(\gamma_\mu\gamma_\nu - \gamma_\nu\gamma_\mu)$ and $\Phi_{\rm v}^{\mu\nu} = \partial^\mu\Phi_{\rm v}^\nu - \partial^\nu\Phi_{\rm v}^\mu$, for π-, η-, ρ-, ω-, δ- and σ-mesons. As in Refs. ([37], [16]), throughout this paper we use the definitions and notations of [6], so, e.g., $\gamma_\mu\gamma_\nu + \gamma_\nu\gamma_\mu = 2g_{\mu\nu}$ ($g_{00} = 1$, $g_{11} = g_{22} = g_{33} = -1$), $\gamma_\mu^\dagger = \gamma_0\gamma_\mu\gamma_0$, $\gamma_5 = i\gamma^0\gamma^1\gamma^2\gamma^3$. In addition, following Ref. [41] we shall distinguish via upper (lower) case letters between the Heisenberg and Dirac picture field operators (Ψ and Φ vs ψ and φ, respectively, for fermions and bosons). Here "bare" constants g_{ps}^0, g_s^0, $g_{\rm v}^0$ and $f_{\rm v}^0$ will be renormalized in the clothing procedure.

In constructing the Hamiltonians with Lagrangian densities (1)-(3) as a departure point, we have first used the equations of motion for the $\boldsymbol{H}$ fields and the Legendre transformation to express the total Hamiltonian H in terms of the independent canonical variables and their conjugates. Then, passing to the D picture (interaction representation) the Hamiltonian has been split into a physically satisfactory free-field part H_0 and an interaction V. Since the component $\Phi_{\rm v}^0$ has no canonical conjugate, we have resorted to a trick prompted by Eq. (7.5.22) from [41] to introduce a proper component $\varphi_{\rm v}^0$ in the Dirac picture. As a result, we arrive to the interaction Hamiltonian densities

$$\mathcal{H}_s(x) = g_s^0\bar{\psi}(x)\psi(x)\varphi_s(x), \tag{4}$$

$$\mathcal{H}_{ps}(x) = ig_{ps}^0\bar{\psi}(x)\gamma_5\psi(x)\varphi_{ps}(x), \tag{5}$$

$$\mathcal{H}_{\rm v}(x) = \mathcal{H}_{\rm sc}(x) + \mathcal{H}_{\rm nsc}(x), \tag{6}$$

where

$$\mathcal{H}_{\rm sc}(x) = g_{\rm v}^0\bar{\psi}(x)\gamma_\mu\psi(x)\varphi_{\rm v}^\mu(x) + \frac{f_{\rm v}^0}{4m^0}\bar{\psi}(x)\sigma_{\mu\nu}\psi(x)\varphi_{\rm v}^{\mu\nu}(x) \tag{7}$$

and

$$\mathcal{H}_{\rm nsc}(x) = \frac{g_{\rm v}^{0\,2}}{2m_{\rm v}^{0\,2}}\bar{\psi}(x)\gamma_0\psi(x)\bar{\psi}(x)\gamma_0\psi(x) + \frac{f_{\rm v}^{0\,2}}{8m^{0\,2}}\bar{\psi}(x)\sigma_{0i}\psi(x)\bar{\psi}(x)\sigma_{0i}\psi(x). \tag{8}$$

It is implied that the total Hamiltonian $H = H_0 + V$ consists of the sum $H_0 = H_{0,f} + \sum_b H_{0,b}$, where $H_{0,f}$ ($H_{0,b}$) is the free-fermion (free-boson) contribution, and the space integral

$$V = \int d\boldsymbol{x}\mathcal{H}(\boldsymbol{x}) \tag{9}$$

of the interaction density $\mathcal{H}(x)$ in the D picture

$$\mathcal{H}(x) = \mathcal{H}_s(x) + \mathcal{H}_{ps}(x) + \mathcal{H}_{\text{v}}(x), \tag{10}$$

taken at $t = 0$, i.e., in the Schrödinger (S) picture. In other words, $\mathcal{H}(\boldsymbol{x}) = \mathcal{H}(0, \boldsymbol{x})$. Expressions (4)-(9) exemplify that for a Lorentz-invariant Lagrangian it is not necessarily to have "... the interaction Hamiltonian as the integral over space of a scalar interaction density; we also need to add nonscalar terms to the interaction density ..." (quoted from p.292 of Ref. [41]). It is the case with derivative couplings and/or spin ≥ 1.

In its turn, the operator

$$H = H_F(\alpha) + H_I(\alpha) \equiv H(\alpha) \tag{11}$$

can be divided into the no-interaction part H_F and the interaction

$$H_I(\alpha) = V(\alpha) + M_{ren}(\alpha) + V_{ren}(\alpha), \tag{12}$$

where free part $H_F(\alpha) \sim \alpha^\dagger \alpha$ belongs to the class [1.1], if one uses the terminology adopted in [37], is a function of creation (destruction) operators $\alpha^\dagger(\alpha)$ in the bare particle representation, i.e., referred to bare particles with physical masses [16]. Here we have introduced notations for the operators $M_{ren}(\alpha) \sim O(g^2)$ ($V_{ren}(\alpha) \sim O(g^3)$) of mass (vertex) counterterms. To be more definite, let us consider interacting fermion and boson fields $V(\alpha) = V_s + V_{ps} + V_{\text{v}}$ with

$$V_s = g_s \int d\boldsymbol{x}\, \bar{\psi}(\boldsymbol{x})\psi(\boldsymbol{x})\varphi_s(\boldsymbol{x}) \tag{13}$$

$$V_{ps} = ig_{ps} \int d\boldsymbol{x}\, \bar{\psi}(\boldsymbol{x})\gamma_5\psi(\boldsymbol{x})\varphi_{ps}(\boldsymbol{x}) \tag{14}$$

$$V_{\text{v}} = V_{\text{v}}^{(1)} + V_{\text{v}}^{(2)} \tag{15}$$

$$V_{\text{v}}^{(1)} = \int d\boldsymbol{x} \left\{ g_{\text{v}}\bar{\psi}(\boldsymbol{x})\gamma_\mu\psi(\boldsymbol{x})\varphi_{\text{v}}^\mu(\boldsymbol{x}) + \frac{f_{\text{v}}}{4m}\bar{\psi}(\boldsymbol{x})\sigma_{\mu\nu}\psi(\boldsymbol{x})\varphi_{\text{v}}^{\mu\nu}(\boldsymbol{x}) \right\} \tag{16}$$

$$V_{\text{v}}^{(2)} = \int d\boldsymbol{x} \left\{ \frac{g_{\text{v}}^2}{2m_{\text{v}}^2}\bar{\psi}(\boldsymbol{x})\gamma_0\psi(\boldsymbol{x})\bar{\psi}(\boldsymbol{x})\gamma_0\psi(\boldsymbol{x}) + \frac{f_{\text{v}}^2}{8m^2}\bar{\psi}(\boldsymbol{x})\sigma_{0i}\psi(\boldsymbol{x})\bar{\psi}(\boldsymbol{x})\sigma_{0i}\psi(\boldsymbol{x}) \right\}, \tag{17}$$

where $\varphi_{\rm V}^{\mu\nu}(\boldsymbol{x}) = \partial^{\mu}\varphi_{\rm V}^{\nu}(\boldsymbol{x}) - \partial^{\nu}\varphi_{\rm V}^{\mu}(\boldsymbol{x})$ the tensor of the vector field included. The mass (vertex) counterterms are given by Eqs. (32)–(33) of Ref. [16] (the difference $V_0(\alpha)$ - $V(\alpha)$ where a primary interaction $V_0(\alpha)$ is derived from $V(\alpha)$ replacing the "physical" coupling constants by the "bare" counterparts).

Moreover, the operator H_I owing to its translation invariance can be written as

$$H_I = \int H_I(\boldsymbol{x})d\boldsymbol{x}, \tag{18}$$

where the interaction density has the property

$$\exp(i\boldsymbol{P}\boldsymbol{a})H_I(\boldsymbol{x})\exp(-i\boldsymbol{P}\boldsymbol{a}) = H_I(\boldsymbol{x}+\boldsymbol{a}) \tag{19}$$

for arbitrary displacement $\boldsymbol{a}$. Our consideration is focused upon various field models (local and nonlocal) in which the interaction density $H_I(\boldsymbol{x})$ consists of scalar $H_{sc}(\boldsymbol{x})$ and nonscalar $H_{nsc}(\boldsymbol{x})$ contributions,

$$H_I(\boldsymbol{x}) = H_{sc}(\boldsymbol{x}) + H_{nsc}(\boldsymbol{x}), \tag{20}$$

where the property to be a scalar means

$$U_F(\Lambda)H_{sc}(x)U_F^{-1}(\Lambda) = H_{sc}(\Lambda x), \quad \forall x = (t, \boldsymbol{x}) \tag{21}$$

for all Lorentz transformations Λ. Henceforth, for any operator $O(\boldsymbol{x})$ in the Schrödinger (S) picture it is introduced its counterpart $O(x) = \exp(iH_F t)O(\boldsymbol{x})\exp(-iH_F t)$ in the Dirac (D) picture.

2.1. Relativistic Field Theory with Particle Creation and Annihilation

In particular, the corresponding set α involves operators $a^{\dagger}(a)$ for the vector bosons, $b^{\dagger}(b)$ for the fermions and $d^{\dagger}(d)$ for the antifermions. Following a common practice, they appear in the standard Fourier expansions

$$\psi(\boldsymbol{x}) = (2\pi)^{-\frac{3}{2}} \int d\boldsymbol{p}\sqrt{\frac{m}{E_{\boldsymbol{p}}}} \sum_{\mu} \left[\bar{u}(\boldsymbol{p}\mu)b(\boldsymbol{p}\mu) + v(-\boldsymbol{p}\mu)d^{\dagger}(-\boldsymbol{p}\mu)\right] e^{i\boldsymbol{p}\boldsymbol{x}}, \tag{22}$$

$$\varphi_{\rm V}^{\mu}(\boldsymbol{x}) = (2\pi)^{-\frac{3}{2}} \int \frac{d\boldsymbol{k}}{\sqrt{2\omega_{\boldsymbol{k}}}} \sum_{s} \left[e^{\mu}(\boldsymbol{k}s)a + e^{\mu}(-\boldsymbol{k}s)a^{\dagger}(-\boldsymbol{k}s)\right] e^{i\boldsymbol{k}\boldsymbol{x}}, \tag{23}$$

where the $e^{\mu}(\boldsymbol{k}s)$ for $s = +1, 0, -1$ are three independent vectors, being transverse $k_{\mu}e^{\mu}(\boldsymbol{k}s) = 0$, and normalized so that

$$\sum_{s} e^{\mu}(\boldsymbol{k}, s)e^{\nu *}(\boldsymbol{k}, s) = -g^{\mu\nu} + k^{\mu}k^{\nu}/m^2 \equiv P_{\rm v}^{\mu\nu}(k). \qquad (24)$$

We will employ the transformation properties of the creation and annihilation operators with respect to Π. For example, in case of a massive particle with the mass m and spin j one considers that

$$U_F(\Lambda, b)a^{\dagger}(p, \mu)U_F^{-1}(\Lambda, b) = e^{i\Lambda pb}D_{\mu'\mu}^{(j)}(W(\Lambda, p))a^{\dagger}(\Lambda p, \mu'), \qquad (25)$$

$$\forall \Lambda \in L_+ \ \text{ and arbitrary spacetime shifts b} = (\mathrm{b}^0, \boldsymbol{b})$$

with D-function whose argument is the Wigner rotation $W(\Lambda, p)$, L_+ the homogeneous (proper) orthochronous Lorentz group. The correspondence $(\Lambda, b) \to U_F(\Lambda, b)$ between elements $(\Lambda, b) \in \Pi$ and unitary transformations $U_F(\Lambda, b)$ realizes an irreducible representation of Π on the Fock space (to be definite) of boson-fermion states. In this context, it is convenient to employ the operators $a(p, \mu) = a(\boldsymbol{p}, \mu)\sqrt{E_{\boldsymbol{p}}}$ that meet the covariant commutation relations

$$\begin{aligned} [a(p'\mu'), a^{\dagger}(p\mu)]_{\pm} &= E_{\boldsymbol{p}}\delta(\boldsymbol{p} - \boldsymbol{p}')\delta_{\mu'\mu}, \\ [a(p'\mu'), a(p\mu)]_{\pm} &= [a^{\dagger}(p'\mu'), a^{\dagger}(p\mu)]_{\pm} = 0. \end{aligned} \qquad (26)$$

Here $E_{\boldsymbol{p}} = \sqrt{\boldsymbol{p}^2 + m^2}$ is the fourth component of the 4-momentum $p = (E_{\boldsymbol{p}}, \boldsymbol{p})$.

Of course, in the particle number representation not only the Hamiltonian $H = \int H(\boldsymbol{x})d\boldsymbol{x}$ is a function of the operators α. Other generators of the Poincaré group Π can be expressed through them, e.g., by means of the so-called Belinfante ansatz [5]

$$\boldsymbol{N} = -\int \boldsymbol{x}H(\boldsymbol{x})d\boldsymbol{x} \qquad (27)$$

for the generator $\boldsymbol{N}$ of the Lorentz boosts, that holds for any local field model with the symmetrized energy-momentum tensor density $\mathcal{T}^{\mu\nu}(x)$. Original approach for constructing the generators of Poincaré group has been exposed in [35, 9, 32].

2.2. The Unitary Clothing Transformation Method in Action

As shown in [37], the Belinfante ansatz turns out to be useful when constructing the Lorentz boosts in the clothed particle representation (CPR). Where generator $\boldsymbol{N} \equiv \boldsymbol{N}(\alpha)$, being determined in the bare particle representation (BPR), is expressed through the new operators $\{\alpha_c\}$ for particle creation and annihilation.

The transition $\{\alpha\} \Longrightarrow \{\alpha_c\}$ is realized via special unitary transformations $W(\alpha) = W(\alpha_c)$, viz.,

$$\alpha = W(\alpha_c)\alpha_c W^\dagger(\alpha_c). \tag{28}$$

These transformations satisfy certain physical requirements:

i) The physical vacuum (the H lowest eigenstate) must coincide with a new no-particle state Ω, i.e., the state that obeys the equations

$$a_c(\boldsymbol{k})\,|\Omega\rangle = b_c(\boldsymbol{p},\mu)\,|\Omega\rangle = d_c(\boldsymbol{p},\mu)\,|\Omega\rangle = 0, \;\; \forall\; \boldsymbol{k},\, \boldsymbol{p},\, \mu \tag{29}$$

$$\langle\Omega|\Omega\rangle = 1.$$

ii) New one-particle states $|\boldsymbol{k}\rangle_c \equiv a_c^\dagger(\boldsymbol{k})\Omega$ etc. are the H eigenvectors as well.

$$K(\alpha_c)|\boldsymbol{k}\rangle_c = K_F(\alpha_c)|\boldsymbol{k}\rangle_c = \omega_k|\boldsymbol{k}\rangle_c \tag{30}$$

$$K_I(\alpha_c)|\boldsymbol{k}\rangle_c = 0 \tag{31}$$

iii) The spectrum of indices that enumerate the new operators must be the same as that for the bare ones .

iv) The new operators α_c satisfy the same commutation rules as do their bare counterparts α that is provided via the link (28) with a unitary operator W to be obtained as in [37].

A key point of the clothing procedure exposed in [37] is to remove the so-called bad terms from the Hamiltonian

$$H \equiv H(\alpha) = H_F(\alpha) + H_I(\alpha) = W(\alpha_c)H(\alpha_c)W^\dagger(\alpha_c) \equiv K(\alpha_c), \tag{32}$$

viz., from a primary interaction $V(\alpha)$ in (12). By definition, such terms prevent the bare vacuum Ω_0 ($a|\Omega_0\rangle = b|\Omega_0\rangle = \ldots = 0$) and the bare one-particle states $|1bare\rangle \equiv a^\dagger|\Omega_0\rangle$ ($b^\dagger|\Omega_0\rangle, \ldots$) to be the H eigenstates.

For example, such terms $b_c^\dagger b_c a_c^\dagger$, $b_c^\dagger d_c^\dagger a_c$, $b_c^\dagger d_c^\dagger a_c^\dagger$, $d_c d_c^\dagger a_c^\dagger$ enter the operator $V(\alpha_c)$ determined by Eqs. (13)–(15) after the replacement of the bare operators in it by the clothed ones. These terms should be removed together with their Hermitian conjugate counterterms to retain the Hermiticity of the similarity transformation (32). In general, the bad terms prevent the physical vacuum $|\Omega\rangle$ (the H lowest eigenstate) and the one-clothed-particle states $|n\rangle_c = a_c^\dagger(n)|\Omega\rangle$, where n the set of quantum numbers necessary for their determination, to be the H eigenvectors for all n included. Here the operators $a_c(n)$ are clothed counterparts of the operators $a(n)$. The bad terms occur every time when any normally ordered product

$$a^\dagger(1')a^\dagger(2')...a^\dagger(n'_C)a(n_A)...a(2)a(1)$$

of the class [C.A] embodies, at least, one substructure which belongs to one of the classes $[k.0]$ ($k = 1, 2, ...$) and $[k.1]$ ($k = 0, 1, ...$).

Keeping in mind the feather CPR application it is convenient to write $V(\alpha) = V_{bad}(\alpha) + V_{good}(\alpha)$, where the term "good", as an antithesis of "bad", is applied here to those operators (e.g., of the class [k.2] with $k \geq 2$) which destroy both the no-clothed-particle state Ω and the one-clothed-particle states. In this respect, all the Yukawa-type couplings in Eqs. (13), (14) and (16) are "bad". Along with the good $a^\dagger a$-, $b^\dagger b$-, $d^\dagger d$-terms in the kinetic energy K_F we meet the terms of such kind in the mass renormalization operators M_{ren}. After this we can rewrite Eq. (32) as

$$\begin{aligned} K(\alpha_c) &= W(\alpha_c)[H_F(\alpha_c) + H_I(\alpha_c)]W^\dagger(\alpha_c) \\ &= W(\alpha_c)[H_F(\alpha_c) + V_{bad}(\alpha_c) + V_{good}(\alpha_c) + M_{ren}(\alpha_c) + V_{ren}(\alpha_c)]W^\dagger(\alpha_c). \end{aligned} \tag{33}$$

or in other form

$$\begin{aligned} K(\alpha_c) = H_F(\alpha_c) + V_{bad}(\alpha_c) + [R, H_F] + [R, V_{bad}] + \frac{1}{2}[R, [R, H_F]] \\ + \frac{1}{2}[R, [R, V_{bad}]] + ... + e^R(V_{good} + M_{ren} + V_{ren})e^{-R} \end{aligned} \tag{34}$$

(cf. Eq. (2.19) in [37]) by requiring

$$[H_F, R] = V_{bad} \tag{35}$$

for the operator R of interest.

One should note that unlike the original clothing procedure exposed in [37], [16] we eliminate here the bad terms only from H_{sc} interaction in spite of such terms can appear in the nonscalar interaction too. This preference is relied upon the previous experience [8] when applying the UCT method in the theory of nucleon-nucleon scattering. Now we get the division

$$H = K(\alpha_c) = K_F + K_I \tag{36}$$

with a new free part $K_F = H_F(\alpha_c) \sim a_c^\dagger a_c$ and interaction

$$K_I = V_{good}(\alpha_c) + M_{ren}(\alpha_c) + V_{ren}(\alpha_c) + [R, V_{good}] + \frac{1}{2}[R, V_{bad}] + [R, M_{ren} + V_{ren}] + \frac{1}{3}[R, [R, V_{bad}]] + ..., \tag{37}$$

where the r.h.s. involves along with good terms other bad terms to be removed via subsequent UCTs described in Sec. 2.4 of [37] and Sec. 3 of [16].

In parallel, we have

$$\boldsymbol{N} \equiv \boldsymbol{N}(\alpha) = \boldsymbol{N}_F(\alpha) + \boldsymbol{N}_I(\alpha) = W(\alpha_c)\boldsymbol{N}(\alpha_c)W^\dagger(\alpha_c) \equiv \boldsymbol{B}(\alpha_c) \tag{38}$$

or

$$\boldsymbol{B}(\alpha_c) = \boldsymbol{N}_F(\alpha_c) + \boldsymbol{N}_I(\alpha_c) + [R, \boldsymbol{N}_F] + [R, \boldsymbol{N}_I] + ..., \tag{39}$$

where accordingly the division

$$\boldsymbol{N}_I = \boldsymbol{N}_B + \boldsymbol{D}, \tag{40}$$

$$\boldsymbol{N}_B = -\int \boldsymbol{x} H_{sc}(\boldsymbol{x}) d\boldsymbol{x} = \boldsymbol{N}_{bad} + \boldsymbol{N}_{good},$$

Eq. (39) can be rewritten as

$$\boldsymbol{B}(\alpha_c) = \boldsymbol{N}_F(\alpha_c) + \boldsymbol{N}_{bad}(\alpha_c) + [R, \boldsymbol{N}_F] + [R, \boldsymbol{N}_{bad}] + \frac{1}{2}[R, [R, \boldsymbol{N}_F]] + \frac{1}{2}[R, [R, \boldsymbol{N}_{bad}]] + ... + e^R \boldsymbol{N}_{good} e^{-R} + e^R \boldsymbol{D} e^{-R}. \tag{41}$$

But it turns out (see the proof of Eq. (3.26) in [37]) that if R meets the condition (35), then

$$[\boldsymbol{N}_F, R] = \boldsymbol{N}_{bad} = -\int \boldsymbol{x} V_{bad}(\boldsymbol{x}) d\boldsymbol{x} \tag{42}$$

so the boost generators in the CPR can be written likely Eq. (36),

$$\boldsymbol{N} = \boldsymbol{B}(\alpha_c) = \boldsymbol{B}_F + \boldsymbol{B}_I, \tag{43}$$

where $\boldsymbol{B}_F = \boldsymbol{N}_F(\alpha_c)$ is the boost operator for noninteracting clothed particles while $\boldsymbol{B}_I$ includes the contributions induced by interactions between them

$$\boldsymbol{B}_I = \boldsymbol{N}_{good}(\alpha_c) + \boldsymbol{D}(\alpha_c) + [R, \boldsymbol{N}_{good}] + \frac{1}{2}[R, \boldsymbol{N}_{bad}] + [R, \boldsymbol{D}] + \frac{1}{3}[R, [R, \boldsymbol{N}_{bad}]] + \ldots \tag{44}$$

One should note that in formulae (37) and (44) we are focused upon the R-commutations with the first-eliminated interaction V_{bad}. As shown in [37], the brackets, on the one hand, yield new interactions responsible for different physical processes and, on the other hand, cancel (as a recipe!) the mass and other counterterms that stem from $H_{nsc}(\alpha_c)$ and $\boldsymbol{D}(\alpha_c)$.

3. Relativistic Interactions between Clothed Particles in Meson-Nucleon Systems

Let us consider the vector bosons (ρ, ω-mesons) interacting with nucleons via the Yukawa-type couplings (15). For brevity, contributions from other mesons are omitted. Then following the scenario, described above, we introduce the first clothing transformation $W^{(1)} = \exp(R_{\rm v}^{(1)})$ ($R_{\rm v}^{(1)\dagger} = -R_{\rm v}^{(1)}$) that eliminates all interactions linear in the coupling constants. Its generator $R_{\rm v}^{(1)}$ obeys the equation

$$\left[R_{\rm v}^{(1)}, H_F\right] + V_{\rm v}^{(1)} = 0 \tag{45}$$

Following [37] Eq. (45) is satisfied with

$$R_{\rm v}^{(1)} = -i \lim_{\varepsilon \to 0+} \int_0^\infty V_D^{(1)}(t) e^{-\varepsilon t} dt \tag{46}$$

if $m_b < 2m$. The corresponding interaction operator in the CPR can be written as

$$K_I^{(2)} = \frac{1}{2}\left[R_{\text{v}}^{(1)}, V_{\text{v}}^{(1)}\right] + V_{\text{v}}^{(2)} + M_{ren}^{(2)}, \tag{47}$$

where we have kept only the contributions of the second order in coupling constants.

Here we do not intend to derive all interactions between the clothed mesons and nucleons that enter the decomposition

$$\begin{aligned} K_I^{(2)} = K(NN \to NN) + K(\bar{N}\bar{N} \to \bar{N}\bar{N}) \\ + K(N\bar{N} \to N\bar{N}) + K(bN \to bN) \\ + K(b\bar{N} \to b\bar{N}) + K(bb' \to N\bar{N}) + K(N\bar{N} \to bb'). \end{aligned} \tag{48}$$

Explicit analytical expressions for the N-N interaction operator in the CPR

$$K(NN \to NN) = \sum_b K_b(NN \to NN), \tag{49}$$

$$\begin{aligned} &K_b(NN \to NN) \\ &= \int \sum_{\mu} d\boldsymbol{p}_1'\, d\boldsymbol{p}_2'\, d\boldsymbol{p}_1\, d\boldsymbol{p}_2 V_b(1', 2'; 1, 2) b_c^\dagger(1') b_c^\dagger(2') b_c(1) b_c(2), \end{aligned} \tag{50}$$

where the symbol $\sum_{\mu}$ denotes the summation over the nucleon spin projections, $1 = \{\boldsymbol{p}_1, \mu_1\}$, etc., have been obtained in [8] (see formulae (19)-(22) therein). In this connection, one should emphasize that while the commutators $\frac{1}{2}[R_s^{(1)}, V_s]$ and $\frac{1}{2}[R_{ps}^{(1)}, V_{ps}]$ generate the scalar- and pseudoscalar-meson contributions $K_b(NN \to NN)$, respectively, in case of the vector mesons we encounter a destructive interplay between the commutator $\frac{1}{2}[R_{\text{v}}^{(1)}, V_{\text{v}}^{(1)}]$ and the integral (17).

To show it explicitly, let us write

$$\begin{aligned} V_{\text{v}}^{(1)} = &-\frac{1}{(2\pi)^{3/2}} \sum_{\mu s} \int d\boldsymbol{k} d\boldsymbol{p}' d\boldsymbol{p} \frac{m}{\sqrt{2\omega_{\boldsymbol{k}} E_{\boldsymbol{p}'} E_{\boldsymbol{p}}}} e^{\rho}(\boldsymbol{k}, s) \delta(\boldsymbol{p}' - \boldsymbol{p} - \boldsymbol{k}) \\ &\times \bar{u}(\boldsymbol{p}'\mu') \left\{ g_{\text{v}}\gamma_\rho - \frac{f_{\text{v}}}{2m} i\sigma_{\nu\rho} k^\nu \right\} u(\boldsymbol{p}\mu) b_c^\dagger(\boldsymbol{p}'\mu') b_c(\boldsymbol{p}\mu) a_c(\boldsymbol{k}, s) + \text{H.c.}, \end{aligned} \tag{51}$$

$$R_{\text{v}}^{(1)} = -\frac{1}{(2\pi)^{3/2}} \sum_{\mu s} \int d\boldsymbol{k} d\boldsymbol{p}' d\boldsymbol{p} \frac{m}{\sqrt{2\omega_{\boldsymbol{k}} E_{\boldsymbol{p}'} E_{\boldsymbol{p}}}} e^{\rho}(\boldsymbol{k}, s) \frac{\delta(\boldsymbol{p}' - \boldsymbol{p} - \boldsymbol{k})}{E_{\boldsymbol{p}'} - E_{\boldsymbol{p}} - \omega_{\boldsymbol{k}}}$$
$$\times \bar{u}(\boldsymbol{p}'\mu') \left\{ g_{\text{v}} \gamma_{\rho} - \frac{f_{\text{v}}}{2m} i\sigma_{\nu\rho} k^{\nu} \right\} u(\boldsymbol{p}\mu) b_c^{\dagger}(\boldsymbol{p}'\mu') b_c(\boldsymbol{p}\mu) a_c(\boldsymbol{k}, s) - \text{H.c.}, \quad (52)$$

retaining only those parts of $V^{(1)}$ and $R^{(1)}$, which are necessary for deriving $K_{\text{v}}(NN \to NN)$. After the normal ordering of the creation (destruction) operators we find

$$\frac{1}{2} \left[R^{(1)}, V^{(1)} \right]_{\text{v}} (NN \to NN) = K_{\text{v}}(NN \to NN) + K_{cont}(NN \to NN) \quad (53)$$

with

$$K_{\text{v}}(NN \to NN) = \int \sum_{\mu} d\boldsymbol{p}_1' \, d\boldsymbol{p}_2' \, d\boldsymbol{p}_1 \, d\boldsymbol{p}_2 V_{\text{v}}(1', 2'; 1, 2) b_c^{\dagger}(1') b_c^{\dagger}(2') b_c(1) b_c(2), \quad (54)$$

$$V_{\text{v}}(1', 2'; 1, 2) = \frac{1}{(2\pi)^3} \frac{m^2}{\sqrt{E_{\boldsymbol{p}_1'} E_{\boldsymbol{p}_2'} E_{\boldsymbol{p}_1} E_{\boldsymbol{p}_2}}} \delta(\boldsymbol{p}_1' + \boldsymbol{p}_2' - \boldsymbol{p}_1 - \boldsymbol{p}_2) \, v_{\text{v}}(1', 2'; 1, 2), \quad (55)$$

$$\begin{aligned} v_{\text{v}}(1', 2'; 1, 2) &= \frac{1}{2} \frac{1}{(p_1' - p_1)^2 - m_{\text{v}}^2} \Bigg[\bar{u}(\boldsymbol{p}_1') \left\{ (g_{\text{v}} + f_{\text{v}}) \gamma_{\nu} - \frac{f_{\text{v}}}{2m} (p_1' + p_1)_{\nu} \right\} u(\boldsymbol{p}_1) \\ &\times \bar{u}(\boldsymbol{p}_2') \left\{ (g_{\text{v}} + f_{\text{v}}) \gamma^{\nu} - \frac{f_{\text{v}}}{2m} (p_2' + p_2)^{\nu} \right\} u(\boldsymbol{p}_2) \\ &- \bar{u}(\boldsymbol{p}_1') \left\{ (g_{\text{v}} + f_{\text{v}}) \gamma_{\nu} - \frac{f_{\text{v}}}{2m} (p_1' + p_1)_{\nu} \right\} u(\boldsymbol{p}_1) \\ &\times \bar{u}(\boldsymbol{p}_2') \frac{f_{\text{v}}}{2m} \left\{ (\hat{p}_1' + \hat{p}_2' - \hat{p}_1 - \hat{p}_2) \gamma^{\nu} - (p_1' + p_2' - p_1 - p_2)^{\nu} \right\} u(\boldsymbol{p}_2) \Bigg], \end{aligned} \quad (56)$$

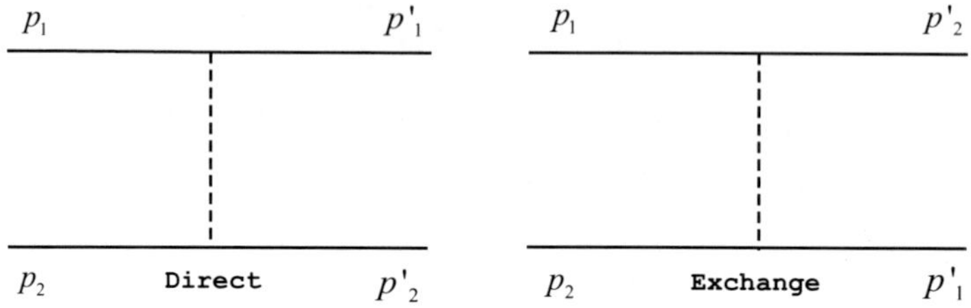

Figure 1. The Feynman-like diagrams for the direct and exchange contributions in the r.h.s. of Eq. (58).

and

$$K_{cont}(NN \to NN) = \frac{1}{2(2\pi)^3} \sum_{\mu} \int d\boldsymbol{p}'_1 d\boldsymbol{p}'_2 d\boldsymbol{p}_1 d\boldsymbol{p}_2 \frac{m^2}{\sqrt{E_{\boldsymbol{p}'_1} E_{\boldsymbol{p}'_2} E_{\boldsymbol{p}_1} E_{\boldsymbol{p}_2}}}$$
$$\times\, \delta\left(\boldsymbol{p}'_1 + \boldsymbol{p}'_2 - \boldsymbol{p}_1 - \boldsymbol{p}_2\right) \left[\frac{g_{\mathrm{v}}^2}{m_{\mathrm{v}}^2} \bar{u}(\boldsymbol{p}'_1\mu'_1)\gamma_0 u(\boldsymbol{p}_1\mu_1)\bar{u}(\boldsymbol{p}'_2\mu'_2)\gamma_0 u(\boldsymbol{p}_2\mu_2)\right.$$
$$\left. - \frac{f_{\mathrm{v}}^2}{4m^2} \bar{u}(\boldsymbol{p}'_1\mu'_1)\gamma_0\boldsymbol{\gamma} u(\boldsymbol{p}_1\mu_1)\bar{u}(\boldsymbol{p}'_2\mu'_2)\gamma_0\boldsymbol{\gamma} u(\boldsymbol{p}_2\mu_2)\right]$$
$$\times\, b_c^\dagger(\boldsymbol{p}'_1\mu'_1) b_c^\dagger(\boldsymbol{p}'_2\mu'_2) b_c(\boldsymbol{p}_1\mu_1) b_c(\boldsymbol{p}_2\mu_2) \quad (57)$$

The latter may be associated with a contact interaction since it does not contain any propagators (cf. the approach by the Osaka group [40]). It is easily seen that this operator cancels completely the nonscalar operator $V^{(2)}$. In other words, the first UCT enables us to remove the non-invariant terms directly in the Hamiltonian. It gives an opportunity to work with the Lorentz scalar interaction only (at least, in the second order in the coupling constants). In our opinion, such a cancellation is a pleasant feature of the CPR.

Further, for each boson included the corresponding relativistic and properly symmetrized $N-N$ interaction (see below), the kernel of integral equations for the $N-N$ bound and scattering states, is determined by

$$\left\langle b_c^\dagger(\boldsymbol{p}'_1) b_c^\dagger(\boldsymbol{p}'_2)\Omega \right| K_b(NN \to NN) \left| b_c^\dagger(\boldsymbol{p}_1) b_c^\dagger(\boldsymbol{p}_2)\Omega \right\rangle$$
$$= V_b^{dir}(1', 2'; 1, 2) - V_b^{exc}(1', 2'; 1, 2), \quad (58)$$

where we have separated the so-called direct

$$V_b^{dir}(1',2';1,2) = -V_b(1',2';1,2) - V_b(2',1';2,1) \tag{59}$$

and exchange

$$V_b^{exc}(1',2';1,2) = V_b^{dir}(2',1';1,2) \tag{60}$$

terms. For example, the one-pion-exchange contribution can be divided into the two parts:

$$V_\pi^{dir}(1',2';1,2) = -\frac{g_\pi^2}{(2\pi)^3}\frac{m^2}{\sqrt{E_{\boldsymbol{p}_1'}E_{\boldsymbol{p}_2'}E_{\boldsymbol{p}_1}E_{\boldsymbol{p}_2}}}\delta\left(\boldsymbol{p}_1'+\boldsymbol{p}_2'-\boldsymbol{p}_1-\boldsymbol{p}_2\right)$$
$$\times\, \bar{u}(\boldsymbol{p}_1')\gamma_5 u(\boldsymbol{p}_1)\bar{u}(\boldsymbol{p}_2')\gamma_5 u(\boldsymbol{p}_2)\frac{1}{2}\left\{\frac{1}{(p_1-p_1')^2-m_\pi^2}+\frac{1}{(p_2-p_2')^2-m_\pi^2}\right\} \tag{61}$$

and

$$V_\pi^{exc}(1',2';1,2) = -\frac{g_\pi^2}{(2\pi)^3}\frac{m^2}{\sqrt{E_{\boldsymbol{p}_1'}E_{\boldsymbol{p}_2'}E_{\boldsymbol{p}_1}E_{\boldsymbol{p}_2}}}\delta\left(\boldsymbol{p}_1'+\boldsymbol{p}_2'-\boldsymbol{p}_1-\boldsymbol{p}_2\right)$$
$$\times\, \bar{u}(\boldsymbol{p}_1')\gamma_5 u(\boldsymbol{p}_2)\bar{u}(\boldsymbol{p}_2')\gamma_5 u(\boldsymbol{p}_1)\frac{1}{2}\left\{\frac{1}{(p_2-p_1')^2-m_\pi^2}+\frac{1}{(p_1-p_2')^2-m_\pi^2}\right\} \tag{62}$$

to be depicted in Fig.1, where the dashed lines correspond to the following Feynman-like "propagators":

$$\frac{1}{2}\left\{\frac{1}{(p_1-p_1')^2-m_\pi^2}+\frac{1}{(p_2-p_2')^2-m_\pi^2}\right\}$$

on the left panel and

$$\frac{1}{2}\left\{\frac{1}{(p_2-p_1')^2-m_\pi^2}+\frac{1}{(p_1-p_2')^2-m_\pi^2}\right\}$$

on the right panel. Note also that expressions (61)-(62) determine the one-pion-exchange part of one-boson-exchange (OBE) interaction derived via the Okubo transformation method in [15] (cf. [19, 16]) taking into account the pion and heavier-meson exchanges.

A distinctive feature of the matrix elements (61)-(62) is presence of covariant Feynman-like "propagators" that are converted into the genuine Feynman propagators on the energy-shell for the N-N scattering, i.e.,

$$E_i \equiv E_{\boldsymbol{p}_1} + E_{\boldsymbol{p}_2} = E_{\boldsymbol{p}_1'} + E_{\boldsymbol{p}_2'} \equiv E_f. \tag{63}$$

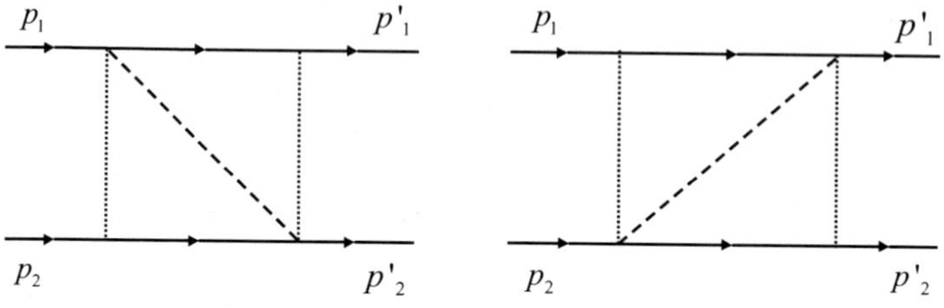

Figure 2. The typical OFPT diagrams with the intermediate boson (dashed lines) on its mass shell.

It is typical of other meson-exchange interactions.

Besides, one should stress that in the course of our derivations the Feynman propagator

$$\left[(p_1 - p_1')^2 - m_b^2\right]^{-1} \tag{64}$$

arises from adding the non-covariant propagators

$$\left[2\omega_{\boldsymbol{k}}\left(E_{\boldsymbol{p}_1} - E_{\boldsymbol{p}_1'} - \omega_{\boldsymbol{k}}\right)\right]^{-1} \tag{65}$$

and

$$\left[2\omega_{\boldsymbol{k}}\left(E_{\boldsymbol{p}_1'} - E_{\boldsymbol{p}_1} - \omega_{\boldsymbol{k}}\right)\right]^{-1}. \tag{66}$$

Such a feature of the UCTs method allows us to use the graphic language of the old-fashioned perturbation theory (OFPT) (see, e.g., Chapter 13 in Schweber's book [29]) when addressing the graphs in Fig.2.

As noted in [16] the graphs within our approach should not be interpreted as the two time-ordered Feynman diagrams. Indeed, all events in the S picture used here are related to the same instant $t = 0$. Being aware of it, the line directions in Fig.2 are given with the sole scope to discriminate between nucleons and antinucleons. The latter will inevitably appear in higher orders in coupling constants and for other physical processes (e.g., the πN scattering) as it has been demonstrated in Ref. [16].

3.1. The Equivalence Theorem: Application to the Neutron-Proton Scattering bellow the Pion Threshold Production

Usually in non-relativistic quantum mechanics (NQM) the LS equation for the T operator

$$T(E+i0) = V_{NN} + V_{NN}(E+i0-h_0)^{-1}T(E+i0) \qquad (67)$$

with a given kernel V_{NN} is a starting point in evaluating the N-N phase shifts. All operators in Eq. (67) (including the sum h_0 of the nucleon kinetic energies) act onto the subspace of two-nucleon states, remaining it invariant (the particle number is conserved within non-relativistic approach). The matrix elements

$$\langle N'N'|T(E+i0)|NN\rangle,$$

make up the corresponding T matrix, the on-energy-shell elements of which at the collision energy

$$E = E_1 + E_2 = E_1^{'} + E_2^{'},$$

can be expressed through the phase shifts and mixing parameters (see below).

In relativistic QFT the situation is completely different. Though one can formally introduce a field operator T that meets the equation

$$T(E+i0) = H_I + H_I(E+i0-H_F)^{-1}T(E+i0) \qquad (68)$$

the field interaction H_I, as a rule, does not conserve the particle number, being the spring of particle creation and destruction. The feature makes the problem of finding the N-N scattering matrix much more complicated than in the framework of the non-relativistic approach since now the T matrix enters an infinite set of coupled integral equations.

Such a general field-theoretic consideration can be simplified with the help of an equivalence theorem [31, 30] according to which the S matrix elements in the Dirac (D) picture, viz.,

$$S_{fi} \equiv \langle \alpha^{\dagger}...\Omega_0|S(\alpha)|\alpha^{\dagger}...\Omega_0\rangle \qquad (69)$$

are equal to the corresponding elements

$$S_{fi}^{c} \equiv \langle \alpha_c^{\dagger}...\Omega|S(\alpha_c)|\alpha_c^{\dagger}...\Omega\rangle \qquad (70)$$

of the S matrix in the CPR. We say "corresponding" keeping in mind the requirement iii). The S operators in Eqs. (69)–(70) are determined by the time evolution from the distant past to the distant future, respectively, for the two decompositions

$$H = H(\alpha) = H_F + H_I$$

and

$$H = K(\alpha_c) = K_F + K_I$$

Note that the equality $S_{fi} = S^c_{fi}$ in question becomes possible owing to certain isomorphism between the α_c algebra and the α algebra once the UCTs $W_D(t) = \exp(iK_F t)W\exp(-iK_F t)$ obey the condition

$$W_D(\pm\infty) = 1 \tag{71}$$

The T operator in the CPR satisfies the equation

$$T_{cloth}(E + i0) = K_I + K_I(E + i0 - K_F)^{-1}T_{cloth}(E + i0) \tag{72}$$

and the matrix

$$T_{fi} \equiv \langle f; b|T(E + i0)|i; b\rangle = \langle f; c|T_{cloth}(E + i0)|i; c\rangle \equiv T^c_{fi}, \tag{73}$$

where $|; b\rangle$ ($|; c\rangle$) are the H_F (K_F) eigenvectors, may be evaluated relying upon properties of the new interaction $K_I(\alpha_c)$.

If in Eq. (72) we approximate K_I by $K_I^{(2)}$ (see Eq. (48)), then initial task of evaluating the BPR matrix elements $\langle N'N'|T(E + i0)|NN\rangle$ can be reduced to solving the equation

$$\langle 1', 2'|T_{NN}(E)|1, 2\rangle = \langle 1', 2'|K_{NN}|1, 2\rangle + \langle 1', 2'|K_{NN}(E + i0 - K_F)^{-1}T_{NN}(E)|1, 2\rangle \tag{74}$$

with $K_{NN} = K(NN \to NN)$.

Here we used the notation $|1, 2\rangle = b_c^\dagger b_c^\dagger|\Omega\rangle$ for any two nucleon state and the completeness condition

$$\sum\!\!\!\!\!\!\!\int_{NN} |NN\rangle\langle NN| = 2, \tag{75}$$

where the symbol $\sum\limits_{NN}$ means the summation over nucleon polarizations and the integration over nucleon momenta.

For practical applications we prefer to work with the corresponding R-matrix,

$$\langle 1'2'|\, R(E)\,|12\rangle = \langle 1'2'|\,\bar{K}_{NN}\,|12\rangle + \sum\!\!\!\!\!\!\int_{34} \langle 1'2'|\,\bar{K}_{NN}\,|34\rangle \frac{\langle 34|\, R(E)\,|12\rangle}{E - E_3 - E_4} \quad (76)$$

with $\bar{K}_{NN} = K_{NN}/2$, where the operation $\sum\!\!\!\!\!\!\int_{34}$ involves the p.v. integration.

As shown in [8], the partial wave expansion of the clothed two-nucleon states, built up in Appendix B of [8], allows us to split Eq. (76) into the set of integral equations

$$R_{l'\,l}^{JST}(p',p) = V_{l'\,l}^{JST}(p',p) + \sum_{l''} \text{p.v.} \int_0^\infty \frac{q^2\,dq}{2(E_p - E_q)} V_{l'\,l''}^{JST}(p',q)\,R_{l''l}^{JST}(q,p) \quad (77)$$

or in other form

$$R_{l'l}^{JST}(p',p) = V_{l'l}^{JST}(p',p) + \frac{1}{2}\sum_{l''} \int_0^\infty \frac{dq}{p^2 - q^2} \{q^2(E_p + E_q)V_{l'l''}^{JST}(p',q)\,R_{l''l}^{JST}(q,p) - 2p^2 E_p V_{l'l''}^{JST}(p',p)\,R_{l''l}^{JST}(p,p)\} \quad (78)$$

to be solved for each submatrix $R_{l'\,l}^{JST}$ composed of half-off-energy shell elements $R_{l'\,l}^{JST}(p',p) \equiv R_{l'\,l}^{JST}(p',p;2E_p)$, where $E_p = \sqrt{\boldsymbol{p}^2 + m^2}$ the collision energy in the center-of-mass system (c.m.s.), m the nucleon mass.

It has to be stressed that unlike the usual NR JST representation the quantum numbers J, S and T in the formula (77) are eigenvalues of the field angular momentum, spin-momentum and isospin operator. Accordingly to Eq. (106) from [8] the common eigenvectors of these operators are determined by[1]

$$|p(lS)JM_J\rangle$$
$$= \int d\hat{\mathbf{p}}\, Y_{lm_l}(\hat{\mathbf{p}})\, (lm_l SM_S|JM_J) \left(\frac{1}{2}\mu_1 \frac{1}{2}\mu_2|SM_S\right) b_c^\dagger(\mathbf{p}\mu_1) b_c^\dagger(-\mathbf{p}\mu_2)\, |\Omega\rangle\,, \tag{79}$$

where the $b_c^\dagger(\mathbf{p}\mu)$ is the creation operator for the clothed nucleon with momentum $\mathbf{p}$ and its polarization μ, $\hat{\mathbf{p}} = \mathbf{p}/p$ the unit vector.

Being aimed at calculations of the low-energy scattering parameters we will recall only the following relations for the on-energy-shell R-matrix elements to the phase shifts and mixing parameters for the uncoupled waves with

$$\tan \delta_J^S = -\pi\rho(p) R_{JJ}^{JS}(p) \tag{80}$$

and for the coupled waves with

$$\tan \delta_\pm^J = -\frac{1}{2}\pi\rho(p) \left[R_{J+1\,J+1}^{J1} + R_{J-1\,J-1}^{J1} \mp \frac{R_{J-1\,J-1}^{J1} - R_{J+1\,J+1}^{J1}}{\cos 2\varepsilon_J} \right] \tag{81}$$

and

$$\tan 2\varepsilon_J = \frac{R_{J+1\,J-1}^{J1} + R_{J-1\,J+1}^{J1}}{R_{J-1\,J-1}^{J1} - R_{J+1\,J+1}^{J1}}, \tag{82}$$

where $\rho(p) = pE_p/2$.

The low-energy scattering parameters are related to the S-wave phase shifts by the following equation

$$p \cot \delta = -\frac{1}{a} + \frac{1}{2} r p^2, \tag{83}$$

where a is the scattering length and r is the effective range.

3.2. Regularized Nucleon-Nucleon Interaction. Description of the Nucleon-Proton Phase-Shifts and Deuteron Properties

Trying to overcome ultraviolet divergences inherent in solving equation (78), we will regularize their driving terms by introducing some cutoff fac-

[1]For brevity, as in [8], we omit isospin indices.

tors. It can be achieved if instead of Eq. (55) one assumes

$$V_b^{reg}(p_1'\mu_1', p_2'\mu_2'; p_1\mu_1, p_2\mu_2) = \frac{\delta(\boldsymbol{p}_1' + \boldsymbol{p}_2' - \boldsymbol{p}_1 - \boldsymbol{p}_2)}{\sqrt{E_{\boldsymbol{p}_1'} E_{\boldsymbol{p}_2'} E_{\boldsymbol{p}_1} E_{\boldsymbol{p}_2}}} v_b^{reg}(p_1'\mu_1', p_2'\mu_2'; p_1\mu_1, p_2\mu_2) \quad (84)$$

omitting for the moment isospin indices, so

$$K_b(NN \to NN) \to K_b^{reg}(NN \to NN) = \int \delta(\boldsymbol{p}_1' + \boldsymbol{p}_2' - \boldsymbol{p}_1 - \boldsymbol{p}_2) \frac{d\boldsymbol{p}_1'}{E_{\boldsymbol{p}_1'}} \frac{d\boldsymbol{p}_2'}{E_{\boldsymbol{p}_2'}} \frac{d\boldsymbol{p}_1}{E_{\boldsymbol{p}_1}} \frac{d\boldsymbol{p}_2}{E_{\boldsymbol{p}_2}}$$
$$\times \sum_{\mu} v_b^{reg}(p_1'\mu_1', p_2'\mu_2'; p_1\mu_1, p_2\mu_2) b_c^\dagger(p_1'\mu_1') b_c^\dagger(p_2'\mu_2') b_c(p_1\mu_1) b_c(p_2\mu_2) \quad (85)$$

Here the new (regularized) coefficients v_b^{reg} are given by

$$v_b^{reg} = F_b(p_1', p_2'; p_1, p_2) v_b, \quad (86)$$

where the old ones v_b are determined by Eqs. (20)–(22) in [8] and empirical cutoff functions F_b [2] should not violate the known symmetries of interactions. In particular, if one writes

$$K_b^{reg}(NN \to NN) = \int K_b^{reg}(\boldsymbol{x}) d\boldsymbol{x} \quad (87)$$

with

$$K_b^{reg}(\boldsymbol{x}) = \frac{1}{(2\pi)^3} \int \frac{d\boldsymbol{p}_1'}{E_{\boldsymbol{p}_1'}} \frac{d\boldsymbol{p}_2'}{E_{\boldsymbol{p}_2'}} \frac{d\boldsymbol{p}_1}{E_{\boldsymbol{p}_1}} \frac{d\boldsymbol{p}_2}{E_{\boldsymbol{p}_2}} d\boldsymbol{x} e^{i(\boldsymbol{p}_1' + \boldsymbol{p}_2' - \boldsymbol{p}_1 - \boldsymbol{p}_2)\boldsymbol{x}}$$
$$\times \sum_{\mu} v_b^{reg}(p_1'\mu_1', p_2'\mu_2'; p_1\mu_1, p_2\mu_2) b_c(p_1'\mu_1') b_c(p_2'\mu_2') b_c^\dagger(p_1\mu_1) b_c^\dagger(p_2\mu_2) \quad (88)$$

then the RI implies the property of the operator

$$K_b^{reg}(x) \equiv \exp(iK_F t)\, K_b^{reg}(\boldsymbol{x}) \exp(-iK_F t) \quad (89)$$

[2]We do not consider the functions depending on nucleon polarizations

to be a scalar, viz.,

$$U_F(\Lambda,\lambda)K_b^{reg}(x)U_F^{-1}(\Lambda,\lambda) = K_b^{reg}(\Lambda x+\lambda) \tag{90}$$

But accordingly (25) it imposes the following restrictions

$$D^{(\frac{1}{2})}_{\eta_1'\mu_1'}(W(\Lambda,p_1'))\,D^{(\frac{1}{2})}_{\eta_2'\mu_2'}(W(\Lambda,p_2'))\,D^{(\frac{1}{2})*}_{\eta_1\mu_1}(W(\Lambda,p_1))\,D^{(\frac{1}{2})*}_{\eta_2\mu_2}(W(\Lambda,p_2))$$
$$\times\, v_b^{reg}(p_1'\mu_1',p_2'\mu_2';p_1\mu_1,p_2\mu_2) = v_b^{reg}(\Lambda p_1'\eta_1',\Lambda p_2'\eta_2';\Lambda p_1\eta_1,\Lambda p_2\eta_2) \tag{91}$$

to the coefficients v_b^{reg}. In this connection, before going on, one needs to verify that the old (non-regularized) ones satisfy relation (91) themselves. It can be done with the help of the property

$$S(\Lambda)u(p\mu) = D^{(1/2)}_{\mu'\mu}(W(\Lambda,p))u(\Lambda p\mu'),$$

where $S(\Lambda) = exp\left[-\frac{1}{2}\omega_{\mu\nu}\sigma^{\mu\nu}\right]$ the matrix of the non-unitary representation $\Lambda \rightarrow S(\Lambda)$ in the space of spinor indices. Indeed, recalling more the relations

$$S(\Lambda)^{-1}\gamma^\rho S(\Lambda) = \gamma^\mu\Lambda^\rho{}_\mu,$$

one can easily seen that the quantities $v_b(p_1'\mu_1',p_2'\mu_2';p_1\mu_1,p_2\mu_2) = v_b(1',2';1,2)$, e.g., $v_{\text{v}}(1',2';1,2)$ defined in Eq. (56), obey Eq. (91). Also, let us remind (see, e.g., [22]) that for a Lorentz boost $\Lambda = L(\boldsymbol{v})$ with the velocity $\boldsymbol{v}$, the Wigner transformation $W(\Lambda,p)$ is the rotation about the $\boldsymbol{v}\times\boldsymbol{p}$ -direction by an angle ψ, which can be represented as

$$m\left(1+\gamma+\frac{p_0^\star}{m}+\frac{p_0}{m}\right)\tan\frac{\psi}{2} = \gamma\left|\boldsymbol{v}\times\boldsymbol{p}\right|,$$

where $p_0^\star$ the zeroth component of the nucleon momentum $p^\star = (p_0^\star,\boldsymbol{p}^\star) = L(\boldsymbol{v})p$ in the moving frame and γ the corresponding Lorentz factor. As noted, $W(R,p) = R$ for a pure rotation R.

Now, keeping in mind the relation (86) we need to deal with a Lorentz-invariant cutoff,

$$F_b(p_1',p_2';p_1,p_2) = F_b(\Lambda p_1',\Lambda p_2';\Lambda p_1,\Lambda p_2) \tag{92}$$

in our model regularization,

$$V_b^{reg}(p_1'\mu_1',p_2'\mu_2';p_1\mu_1,p_2\mu_2) = \frac{\delta\left(\boldsymbol{p}_1'+\boldsymbol{p}_2'-\boldsymbol{p}_1-\boldsymbol{p}_2\right)}{\sqrt{E_{\boldsymbol{p}_1'}E_{\boldsymbol{p}_2'}E_{\boldsymbol{p}_1}E_{\boldsymbol{p}_2}}}$$
$$F_b(p_1',p_2';p_1,p_2)v_b^{reg}(p_1'\mu_1',p_2'\mu_2';p_1\mu_1,p_2\mu_2) \tag{93}$$

In order to facilitate comparison with some derivations and calculations from Refs. [21], [20], we employ the notation

$$\langle \boldsymbol{p}'\,\mu_1'\mu_2' |\, v_b^{UCT}\, | \boldsymbol{p}\,\mu_1\mu_2 \rangle \equiv -F_b^2(p',p)b\left[v_b(1',2';1,2)+v_b(2',1';2,1)\right]$$

for the regularized UCT quasipotentials in the c.m.s. (see Appendix C in [8]). As in Ref.[21], we put

$$F_b(p',p)=\left[\frac{\Lambda_b^2-m_b^2}{\Lambda_b^2-(p'-p)^2}\right]^{n_b}\equiv F_b[(p'-p)^2].$$

Doing so, we have

$$\begin{aligned}
&\langle \boldsymbol{p}'\,\mu_1'\mu_2' |\, v_s^{UCT}\, | \boldsymbol{p}\,\mu_1\mu_2 \rangle \\
&\qquad = g_s^2\bar{u}(\boldsymbol{p}')u(\boldsymbol{p})\frac{F_s^2[(p'-p)^2]}{(p'-p)^2-m_s^2}\bar{u}(-\boldsymbol{p}')u(-\boldsymbol{p}),
\end{aligned} \tag{94}$$

$$\begin{aligned}
&\langle \boldsymbol{p}'\,\mu_1'\mu_2' |\, v_{ps}^{UCT}\, | \boldsymbol{p}\,\mu_1\mu_2 \rangle \\
&\qquad = -g_{ps}^2\bar{u}(\boldsymbol{p}')\gamma_5 u(\boldsymbol{p})\frac{F_{ps}^2[(p'-p)^2]}{(p'-p)^2-m_{ps}^2}\bar{u}(-\boldsymbol{p}')\gamma_5 u(-\boldsymbol{p})
\end{aligned} \tag{95}$$

and

$$\begin{aligned}
&\langle \boldsymbol{p}'\,\mu_1'\mu_2' |\, v_{\mathrm{v}}^{UCT}\, | \boldsymbol{p}\,\mu_1\mu_2 \rangle = -\frac{F_{\mathrm{v}}^2[(p'-p)^2]}{(p'-p)^2-m_{\mathrm{v}}^2} \\
&\times\left\{\bar{u}(\boldsymbol{p}')\left[(g_{\mathrm{v}}+f_{\mathrm{v}})\gamma_\nu-\frac{f_{\mathrm{v}}}{2m}(p'+p)_\nu-\frac{f_{\mathrm{v}}}{2m}(E_{\boldsymbol{p}'}-E_{\boldsymbol{p}})[\gamma_0\gamma_\nu-g_{0\nu}]\right]u(\boldsymbol{p})\right. \\
&\times\bar{u}(-\boldsymbol{p}')\left[(g_{\mathrm{v}}+f_{\mathrm{v}})\gamma^\nu-\frac{f_{\mathrm{v}}}{2m}\overline{(p'+p)}^\nu-\frac{f_{\mathrm{v}}}{2m}(E_{\boldsymbol{p}'}-E_{\boldsymbol{p}})[\gamma^0\gamma^\nu-g^{0\nu}]\right]u(-\boldsymbol{p}) \\
&\left.-\frac{f_{\mathrm{v}}{}^2}{4m^2}(E_{p'}-E_p)^2\bar{u}(\boldsymbol{p}')[\gamma_0\gamma_\nu-g_{0\nu}]u(\boldsymbol{p})\bar{u}(-\boldsymbol{p}')[\gamma^0\gamma^\nu-g^{0\nu}]u(-\boldsymbol{p})\right\},
\end{aligned} \tag{96}$$

where $\overline{(p'+p)}^\nu=(E_{\boldsymbol{p}'}+E_{\boldsymbol{p}},-(\boldsymbol{p}'+\boldsymbol{p}))$. At the first sight, such a regularization can be achieved via a simple substitution $g_b\rightarrow g_bF_b(p',p)$ with some cutoff functions $F_b(p',p)$ depending on the 4-momenta p' and p. However, a principal moment is to satisfy the requirement (90) for the Hamiltonian invariant under space inversion, time reversal and charge conjugation. A constructive consideration of the issue is given in Appendix C of Ref. [8].

Replacing in equations (94)–(96)

$$F_b^2[(p'-p)^2]\{(p'-p)^2-m_b^2\}^{-1}$$

by

$$-F_b^2[-(\boldsymbol{p}'-\boldsymbol{p})^2]\{(\boldsymbol{p}'-\boldsymbol{p})^2+m_b^2\}^{-1}$$

and neglecting the tensor-tensor term

$$\frac{f_{\mathrm{v}}{}^2}{4m^2}(E_{p'}-E_p)^2\bar{u}(\boldsymbol{p}')[\gamma_0\gamma_\nu-g_{0\nu}]u(\boldsymbol{p})\bar{u}(-\boldsymbol{p}')[\gamma^0\gamma^\nu-g^{0\nu}]u(-\boldsymbol{p}) \quad (97)$$

in (96), we obtain approximate expressions that with the common factor

$$(2\pi)^{-3}m^2/E_{p'}E_p$$

instead of

$$(2\pi)^{-3}m/\sqrt{E_{p'}E_p}$$

are equivalent to Eqs. (E.21)–(E.23) from [21]. Such an equivalence becomes coincidence if in our formulae instead of the canonical two-nucleon basis $|\boldsymbol{p}\,\mu_1\mu_2\rangle$ one uses the helicity basis as in [21].

In parallel, we have considered the set of equations

$$\begin{aligned}{}^B R_{l'l}^{JST}(p',p) &= {}^B V_{l'l}^{JST}(p',p)\\ &+ m\sum_{l''}\int_0^\infty \frac{dq}{p^2-q^2}\left\{q^2\,{}^B V_{l'l''}^{JST}(p',q)\,{}^B R_{l''l}^{JST}(q,p)\right.\\ &\left.-p^2\,{}^B V_{l'l''}^{JST}(p',p)\,{}^B R_{l''l}^{JST}(p,p)\right\}, \end{aligned} \quad (98)$$

where the superscript B refers to the partial matrix elements of the potential B determined in [20] with the just mentioned interchange of the bases. It is important to note that Eqs. (98) can be obtained from Eqs. (78) ignoring some relativistic effects. In particular, it means that the covariant OBE propagators

$$\frac{1}{(p'-p)^2-m_b^2}=\frac{1}{(E_{p'}-E_p)^2-(\boldsymbol{p}'-\boldsymbol{p})^2-m_b^2} \quad (99)$$

are replaced by their non-relativistic counterparts

$$-\frac{1}{(\boldsymbol{p}'-\boldsymbol{p})^2+m_b^2}. \quad (100)$$

Such an approximation [3] is a key point that gives rise to the potential B from [20].

After these preliminaries, we will present several results obtained within this approach.

Table 1. The best-fit parameters for the Bonn and Kharkov potentials

Meson		Bonn B	UCT Bonn	UCT GS
π	$g_\pi^2/4\pi$	14.4	14.633	14.3868
	Λ_π	1700	2330.4317	2316.5957
	m_π	138.03	138.03	138.03
η	$g_\eta^2/4\pi$	3	3.8712	4.7436
	Λ_η	1500	1148.3563	1186.3328
	m_η	548.8	548.8	548.8
ρ	$g_\rho^2/4\pi$	0.9	1.5239	1.4905
	Λ_ρ	1850	1470.0933	1482.9515
	f_ρ/g_ρ	6.1	5.4099	5.63504
	m_ρ	769	769	769
ω	$g_\omega^2/4\pi$	24.5	27.0059	27.0010
	Λ_ω	1850	2067.1625	2048.4847
	m_ω	782.6	782.6	782.6
δ	$g_\delta^2/4\pi$	2.488	1.8362	1.9911
	Λ_δ	2000	2283.0762	2117.1415
	m_δ	983	983	983
σ, $T=0$ ($T=1$)	$g_\sigma^2/4\pi$	18.3773 (8.9437)	18.8026 (10.7836)	18.9937 (10.8998)
	Λ_σ	2000 (1900)	1629.1474 (2123.1678)	1738.8244 (2145.0415)
	m_σ	720 (550)	722.22 (565.79)	723.64 (571.74)

First of all, we show (Table 1) the optimum values of the adjustable parameters extracted from the phase-shift analyses of the neutron-proton elastic scattering at the collision energies up to the pion production threshold. The third column taken from table A.1. in [20], the forth (fifth) column has been obtained via fitting the solutions of the integral equations for the R-matrix elements from [8] to the Bonn phase-shifts [20] (WCJ1 ones [11]). All masses are in MeV. One should stress that corresponding time-consuming fit procedure has been essentially reduced by using a special code elaborated in the Computing Center of Saint-Petersburg Uni-

[3]Sometimes associated with ignoring the so-called meson retardation (see, e.g., Appendix E from [21] and a discussion therein)

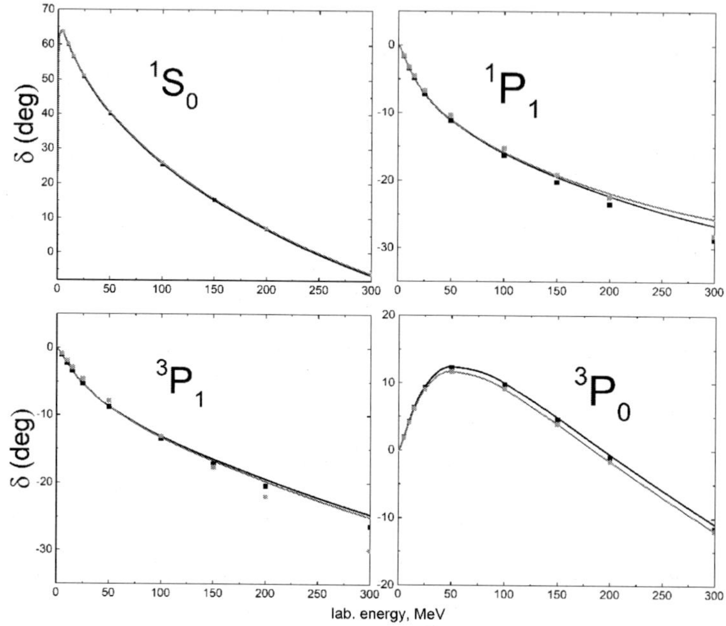

Figure 3. Neutron-proton phase-shifts for the uncoupled partial waves versus the nucleon kinetic energy in the lab. frame. Black (red) curves for the Kharkov potential with the UCT Bonn (UCT GS) parameters from Table 1. Black (green) points for the Bonn B (WCJ1) potential.

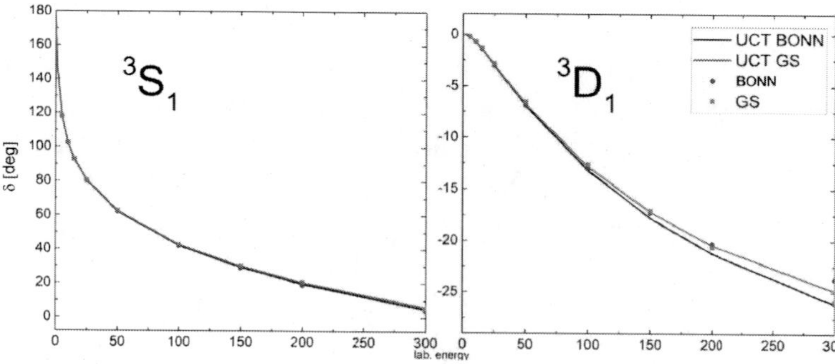

Figure 4. The same as in Figure 3 but for the first coupled waves.

Table 2. Low-energy parameters and deuteron binding energy. The experimental values are from Table 4.2 of Ref. [21]

Parameter	Bonn B	WCJ1	UCT Bonn	UCT GS	Experiment
a_s (fm)	-23.685	-23.749	-23.695	-23.728	-23.748±0.010
r_s (fm)	2.71	2.67	2.71	2.69	2.68±0.05
a_t (fm)	5.426	5.429	5.431	5.421	5.419±0.007
r_t (fm)	1.761	1.766	1.769	1.755	1.754±0.008
ε_d (MeV)	2.222	2.222	2.223	2.222	2.224575

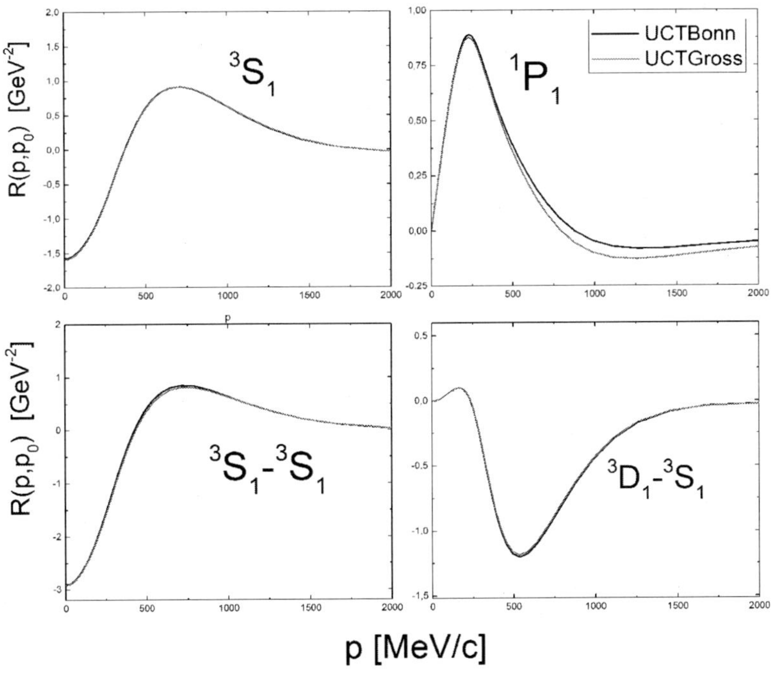

Figure 5. Half-off-shell R-matrices for uncoupled waves at lab. energy equal to 150 MeV ($p_0 = 265$MeV). The difference between the curves are the same as in Fig.3.

versity, Russia. Just such best-fit values determine the two versions of the so-called Kharkov potential [12, 2].

On-shell calculations below the pion production threshold are shown in Figs.3-4. As seen in Fig.3 the UCT GS fit ensures a slightly better treatment of the phase-shifts in ${}^1P_1$ and ${}^3P_0$ states. The same refers to the agreement between our predictions of the energy dependence of the ${}^3D_1$ phase-shift in Fig.4. Some off-shell effects are demonstrated in Fig.5 with the off-energy-shell R-matrices $R(p', p_0)$ for first partial waves. Recall that on-shell R-matrix elements $R(p_0, p_0)$ are proportional to $tan\delta(p_0)$.

As the by-products of our calculations the corresponding values of the low-energy parameters for the elastic neutron-proton scattering are collected in Table 2 together with the deuteron binding energies for the two model potentials. We see that the UCT GS values are closer to the data than the UCT Bonn ones.

At this point we allow ourselves recall a possible way [8] when finding the deuteron state $|\Psi_d(\boldsymbol{P})\rangle \in \mathcal{H}_{2N}$ that meets the eigenvalue equation

$$[H_F(\alpha) + H_I(\alpha)]|\Psi_d(\boldsymbol{P})\rangle = E_d|\Psi_d(\boldsymbol{P})\rangle \tag{101}$$

or in the CPR

$$[K_F(\alpha) + K_I(\alpha_c)]|\Psi_d(\boldsymbol{P})\rangle = E_d|\Psi_d(\boldsymbol{P})\rangle, \tag{102}$$

with $E_d = \sqrt{m_d^2 + \boldsymbol{P}^2}$, where $\boldsymbol{P}$ is the total deuteron momentum, $m_d = m_p + m_n - \varepsilon_d$ is the deuteron mass and ε_d represents the binding energy of the deuteron.

Using the approximation with $K_I(\alpha_c) = K_{NN}$ we arrive to a simpler eigenvalue problem

$$\left[K_2^N + K_{NN}\right]|\boldsymbol{P}; M\rangle = E_d|\boldsymbol{P}; M\rangle \tag{103}$$

in the subspace $\mathcal{H}_{2N}$ spanned onto the basis $b_c^\dagger b_c^\dagger|\Omega\rangle$ with $K_{NN} \sim b_c^\dagger b_c^\dagger b_c b_c$. Here M denotes the deuteron spin projection on quantization axis. The solution of this equation can be represented as[4]

$$|\boldsymbol{P}; M\rangle = \int d\boldsymbol{p}_1 d\boldsymbol{p}_2 D_M(\boldsymbol{P}; \boldsymbol{p}_1\mu_1, \boldsymbol{p}_2\mu_2) b_c^\dagger(\boldsymbol{p}_1\mu_1) b_c^\dagger(\boldsymbol{p}_2\mu_2)|\Omega\rangle, \tag{104}$$

with the c-number coefficients $D_M(\boldsymbol{P}; \boldsymbol{p}_1\mu_1, \boldsymbol{p}_2\mu_2) = \delta(\boldsymbol{P} - \boldsymbol{p}_1 - \boldsymbol{p}_2)\psi_M(\boldsymbol{p}_1\mu_1, \boldsymbol{p}_2\mu_2)$ that have the property

$$\psi_M(1,2) = -\psi_M(2,1). \tag{105}$$

[4]For a moment, the isospin quantum numbers are suppressed.

Sometimes it is convenient to write $\{\boldsymbol{p}_1, \mu_1\} = 1$, etc.

In the deuteron rest frame the equation (103) takes the form

$$\left[K_2^N + K_{NN}\right] |\psi_M\rangle = m_d |\psi_M\rangle \tag{106}$$

or

$$|\psi_M\rangle = \left[m_d - K_2^N\right]^{-1} K_{NN} |\psi_M\rangle, \tag{107}$$

where

$$|\psi_M\rangle \equiv |\boldsymbol{P} = 0; M\rangle = \int d\boldsymbol{p} \psi_M(\boldsymbol{p}\mu_1, -\boldsymbol{p}\mu_2) b_c^\dagger(\boldsymbol{p}\mu_1) b_c^\dagger(-\boldsymbol{p}\mu_2)|\Omega\rangle. \tag{108}$$

Using the basis vectors $|p(lS)JM_J, TM_T\rangle$ introduced in our previous paper [34] (see Appendix B) the vector $|\psi_M\rangle$ can be written as

$$|\psi_{M,TM_T}\rangle = \frac{1}{\sqrt{2}} \sum \int_0^\infty p^2 dp\, |p(lS)1M, TM_T\rangle \psi_{lST}(p), \tag{109}$$

since the deuteron has the invariant spin equal to $J = 1$. In Eq. (109) the permissible values of the quantum numbers l, S and T are restricted to the property

$$\mathcal{P}_{ferm} |\psi_{M,TM_T}\rangle = |\psi_{M,TM_T}\rangle, \tag{110}$$

with respect to the space inversion (see Appendix B in [8], where one can find formula (114) for the parity operator $\mathcal{P}_{ferm}$ of the nucleon field in the CPR). In fact, there are only the two combinations of T, S and l, namely, $T = 0$, $S = 1$ and $l = 0, 2$. Respectively,

$$|\psi_{M,00}\rangle \equiv |\psi_M\rangle = \frac{1}{\sqrt{2}} \sum_{l=0,2} \int_0^\infty p^2 dp\, |p(l1)1M\rangle \psi_l(p). \tag{111}$$

Combining Eqs. (108) and (111), we obtain

$$\begin{aligned} &\psi_M(\boldsymbol{p}\mu_1\tau_1, -\boldsymbol{p}\mu_2\tau_2) \\ &= \frac{1}{\sqrt{2}} \sum \psi_l(p) Y_{lm_l}(\hat{\boldsymbol{p}}) (lm_l 1 M_S|1M) \left(\tfrac{1}{2}\mu_1 \tfrac{1}{2}\mu_2 | S M_S\right) \left(\tfrac{1}{2}\tau_1 \tfrac{1}{2}\tau_2 | 00\right) \end{aligned} \tag{112}$$

with the property

$$\psi_M(\boldsymbol{p}\mu_1\tau_1, -\boldsymbol{p}\mu_2\tau_2) = -\psi_M(-\boldsymbol{p}\mu_2\tau_2, \boldsymbol{p}\mu_1\tau_1) \tag{113}$$

which will be used later. Here $\tau_i (i = 1, 2)$ are nucleon isospin projections. At this point, we accept the normalization condition [5]

$$\langle\psi_{M'}|\psi_M\rangle = \delta_{M'M} \tag{114}$$

which is equivalent to

$$2\sum \int d\boldsymbol{p}\psi^*_{M'}(\boldsymbol{p}\mu_1, -\boldsymbol{p}\mu_2)\psi_M(\boldsymbol{p}\mu_1, -\boldsymbol{p}\mu_2) = \delta_{M'M}, \tag{115}$$

that implies

$$\int\limits_0^\infty p^2 dp \left[\psi_0^2(p) + \psi_2^2(p)\right] = 1. \tag{116}$$

Substituting the decomposition (111) into the equation (107) we get the set of homogeneous integral equations for "radial" components $\psi_l(p)(l = 0, 2)$

$$\psi_l(p) = \frac{1}{m_d - 2E_{\boldsymbol{p}}} \sum_{l'} \int\limits_0^\infty k^2 dk V_{l\,l'}^{J=S=1,T=0}(p, k)\psi_{l'}(k). \tag{117}$$

In a moving frame the corresponding eigenvector that belongs to the value $E_d = \sqrt{\boldsymbol{P}^2 + m_d^2}$ can be determined either by solving directly the equation (103) or using the relation

$$|\boldsymbol{P}; M\rangle = \exp[-i\boldsymbol{\beta}\boldsymbol{B}(\alpha_c)]|\psi_M\rangle \tag{118}$$

The boost operator $\boldsymbol{B}(\alpha_c) = \boldsymbol{B}_F(\alpha_c) + \boldsymbol{B}_I(\alpha_c)$ determined in the CPR by

$$\boldsymbol{B}(\alpha_c) = W(\alpha_c)\boldsymbol{N}(\alpha_c)W^\dagger(\alpha_c), \tag{119}$$

[5] More precisely, relation (103) should be formulated for the wave packets $\int d\boldsymbol{P}a(\boldsymbol{P})|\boldsymbol{P}; M\rangle$. Each of them is a superposition of eigenvectors of the total momentum operator with eigenvalues $\boldsymbol{P}$ close to $\boldsymbol{P} = 0$ by letting the packet width goes to zero at the end of calculations.

consists of the free $\boldsymbol{B}_F$ and interaction $\boldsymbol{B}_I$ parts. Here $\boldsymbol{N}$ is the total boost operator for interacting fields.

Perhaps, one should note that the required

$$\hat{P}^{\mu}|\boldsymbol{P}; M\rangle = P^{\mu}|\boldsymbol{P}; M\rangle \tag{120}$$

follows from the property of the energy-momentum operator $\hat{P}^{\mu} = (H, \hat{P}^1, \hat{P}^2, \hat{P}^3)$ to be the four-vector, viz.,

$$e^{-i\boldsymbol{\beta}\boldsymbol{B}}\hat{P}^{\mu}e^{i\boldsymbol{\beta}\boldsymbol{B}} = \hat{P}^{\nu}L_{\nu}^{\ \mu}(\boldsymbol{\beta}) \tag{121}$$

with the matrix

$$L(\boldsymbol{\beta}) = \begin{bmatrix} L_0^{\ 0} = P^0/m_d & \vdots & L_0^{\ j} = P^j/m_d \\ \cdots & \cdots & \cdots \\ L_i^{\ 0} = P_i/m_d & \vdots & L_i^{\ j} = \delta_i^j - \dfrac{P_i P^j}{m_d(P^0 + m_d)} \end{bmatrix} \tag{122}$$

of the Lorentz transformation $m_d(1,0,0,0) \Rightarrow (P^0, P^1, P^2, P^3) = P$. In these formulae the parameters $(\beta^1, \beta^2, \beta^3) = \boldsymbol{\beta}$ are related to the velocity $\boldsymbol{v} = \boldsymbol{P}/m_d$ of the moving frame

$$\boldsymbol{\beta} = \beta\boldsymbol{n}, \quad \boldsymbol{n} = \boldsymbol{v}/v, \quad \tanh\beta = v. \tag{123}$$

In order to meet Eq. (121) it is sufficient to have the two Lie-Poincaré commutations

$$[K, \boldsymbol{B}] = i\boldsymbol{P}, \quad [P_i, B_j] = i\delta_{ij}K. \tag{124}$$

To the approximation that leads to $K_I = K(NN \rightarrow NN)$ the operator $B_I \sim b_c^\dagger b_c^\dagger b_c b_c$, i.e., it repeats the operator structure of K_I [35]. Then it is readily seen (cf., derivations below Eq. (38) in [8]) that the vector (118) with $\boldsymbol{P} \neq 0$ belongs to the sector $\mathcal{H}_{2N}$, i.e., it has the structure (104).

In addition, we have obtained the following values for low-energy parameters and deuteron binding energy in Table 2. We see that the UCT GS values are closer to the experimental points than the UCT Bonn ones.

At last, we have calculated in Fig.6 the p-dependencies of the deuteron wave function components to distinguish between not only the Bonn and Kharkov potential predictions but also between the UCT Bonn and UCT GS versions of the latter. The preceding experience prompts us that such

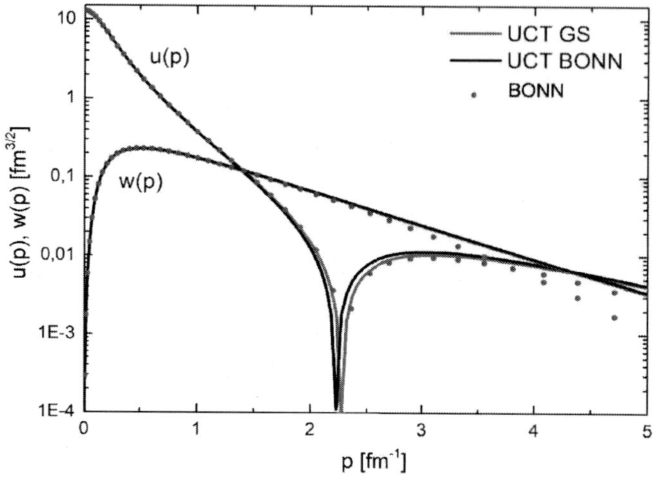

Figure 6. Deuteron wave functions for $\psi_0^d(p) = u(p)$ and $\psi_2^d(p) = w(p)$. Black (red) curves for Kharkov potential with the parameters of UCT Bonn (UCT GS) fit. Points for Bonn B potential.

distinctions can be revealed when studying the proton polarization in the deuteron breakup $d(\boldsymbol{e}, e'\boldsymbol{p})n$ [22] and the target asymmetry and azimuthal asymmetries in the electrodisintegration of polarized deuterons $\boldsymbol{d}(e, e'p)n$ [23]. Again, we note a better treatment of the experimental data via the UCT GS version.

3.3. The Clothed Particle Representation in the Theory of $3N$-Systems

Like the 2-body case [37, 34] the $3N$ eigenvalue problem in the CPR can be formulated in the following way (cf. Eq. (102))

$$\left[K_2^N + K_I\right] |\Psi\rangle = E|\Psi\rangle, \tag{125}$$

where the state $|\Psi\rangle$ belongs to the $3N$ sector of the Fock $\mathcal{H}_{3N}$ that spanned onto the basis $b_c^\dagger b_c^\dagger b_c^\dagger|\Omega\rangle$. Here interaction operator consists of two terms $K_I = K_{NN} + K_{NNN}$, where besides the nucleon-nucleon interaction $K_{NN} \sim b_c^\dagger b_c^\dagger b_c b_c$ one has to deal with the $3N$-force interaction operator $K_{NNN} \sim b_c^\dagger b_c^\dagger b_c^\dagger b_c b_c b_c$.

Similarly to the deuteron, the $3N$ bound state (triton) is written as a superposition of the basis states in $\mathcal{H}_{3N}$

$$|\Psi(\boldsymbol{P})\rangle = \int d\boldsymbol{p}_1 d\boldsymbol{p}_2 d\boldsymbol{p}_3 \Psi_{\boldsymbol{P}}(\boldsymbol{p}_1, \boldsymbol{p}_2, \boldsymbol{p}_3) b_c^\dagger(\boldsymbol{p}_1) b_c^\dagger(\boldsymbol{p}_2) b_c^\dagger(\boldsymbol{p}_3) \tag{126}$$

with the c-number function $\Psi_{\boldsymbol{P}}(\boldsymbol{p}_1, \boldsymbol{p}_2, \boldsymbol{p}_3) = \delta(\boldsymbol{P} - \boldsymbol{p}_1 - \boldsymbol{p}_2 - \boldsymbol{p}_3)\psi_T(\boldsymbol{p}_1, \boldsymbol{p}_2, \boldsymbol{p}_3)$. To obtain equation for this function one needs to project the equation (125) onto basis states $|123\rangle = b_c^\dagger(\boldsymbol{p}_1) b_c^\dagger(\boldsymbol{p}_2) b_c^\dagger(\boldsymbol{p}_3)|\Omega\rangle$ as

$$\langle 123|[K_2^N + K_I]|\Psi(\boldsymbol{P})\rangle = E\Psi_{\boldsymbol{P}}(\boldsymbol{p}_1, \boldsymbol{p}_2, \boldsymbol{p}_3), \tag{127}$$

and taking into account that the operators b_c destroy the physical vacuum $|\Omega\rangle$, *i.e.*, $b_c|\Omega\rangle = 0$. Omitting the $3N$-forces in K_I, we obtain

$$(E_1 + E_2 + E_3)\Psi_{\boldsymbol{P}}(1,2,3) + \langle 123|V_1 + V_2 + V_3|\Psi(\boldsymbol{P})\rangle = E\Psi_{\boldsymbol{P}}(1,2,3), \tag{128}$$

with $E_i = \sqrt{\boldsymbol{p}_i^2 + m^2}$, $V_1 = \tilde{V}_{NN}(2,3)$, $V_2 = \tilde{V}_{NN}(1,3)$, $V_3 = \tilde{V}_{NN}(1,2)$ and

$$\begin{aligned}&\tilde{V}_{NN}(i,j)\\ &= -V_{NN}(i'\,j';i\,j) + V_{NN}(i'\,j';j\,i) - V_{NN}(j'\,i';j\,i) + V_{NN}(j'\,i';i\,j)\,.\end{aligned} \tag{129}$$

where nucleon-nucleon quasipotentials V_{NN} is given by

$$V_{NN}(i',j',i,j) = \sum_b V_b^{reg}(i',j';i,j), \tag{130}$$

with V_b^{reg} from Eq. (84).

Equation (128) can be viewed as a relativistic version of the Faddeev equation in the CPR with account of pair interactions. Of course, the $3N$-forces should be included for a more complete description. To derive the corresponding interaction operator K_{NNN} one needs to evaluate more complicated R-commutators, e.g., in case of the pion exchange, where the commutator $[R[R[R, V_{ps}]]]$ is a departure point in constructing the $3N$-force of interest. Therefore, in the CPR one can obtain the interaction operators for the $2N$-forces and $3N$-forces in a consistent way. As an example, we will confine ourselves to the forces generated by the pion exchange merely.

3.4. The 2N-, 3N- and More Complicated Interaction Operators between Clothed Nucleons

In this context, following [16] we consider the multiple commutators

$$[V]^n = [R, [R, ...[R, V]...]], \tag{131}$$

with n-brackets ($n = 1, 2, ...$). In the framework of the Yukawa model or any other field model with a polynomial interaction the operator V can be represented in the following symbolic form

$$V \equiv f * m + H.c., \tag{132}$$

where $f * m$ is a polynomial composed of products of fermionic and mesonic operators.

In general, to obtain recursive relations for the commutators of increasing complexity, it is convenient to write down

$$[V]^n = \lim_{\lambda \to 0} \frac{d^n}{d\lambda^n} \left(e^{\lambda R} f * m e^{-\lambda R} \right) + H.c.. \tag{133}$$

or

$$[V]^n = \lim_{\lambda \to 0} \frac{d^n}{d\lambda^n} (f(\lambda) * m(\lambda)) + H.c.. \tag{134}$$

By definition, $f(\lambda) \equiv e^{\lambda R} f e^{-\lambda R}$ and $m(\lambda) \equiv e^{\lambda R} m e^{-\lambda R}$.

Then, using the Leibnitz formula

$$\frac{d^n}{d\lambda^n} (f(\lambda) * m(\lambda)) = \sum_{s=0}^{n} C_n^s [f(\lambda)]^{n-s} * [m(\lambda)]^s,$$

we find

$$\lim_{\lambda \to 0} \frac{d^n}{d\lambda^n} (f(\lambda) * m(\lambda)) = \sum_{s=0}^{n} C_n^s [f(0)]^{n-s} * [m(0)]^s,$$

whence

$$[V_{bad}]^n = \sum_{s=0}^{n} C_n^s [f]^{n-s} * [m]^s + H.c.. \tag{135}$$

At the first stage of our procedure for the Yukawa model

$$V_{bad} = V(\alpha_c) = \int d\boldsymbol{k} f(\boldsymbol{k}) m(\boldsymbol{k}) + H.c.,$$
$$R = R_1(\alpha_c) = \int d\boldsymbol{k} \hat{R}_c^{\boldsymbol{k}} m(\boldsymbol{k}) - H.c., \quad (136)$$

with $f(\boldsymbol{k}) = \hat{V}_c^{\boldsymbol{k}}$ and $m(\boldsymbol{k}) = a_c(\boldsymbol{k})$. For such V_{bad} formula (135), where $[f]^{n-s} * [m]^s \equiv \int d\boldsymbol{k}[f(\boldsymbol{k})]^{n-s}[m(\boldsymbol{k})]^s$, yields

$$[V_{bad}]^1 = \int d\boldsymbol{k} \left[[f(\boldsymbol{k})]^1 m(\boldsymbol{k}) + f(\boldsymbol{k})[m(\boldsymbol{k})]^1 \right] + H.c., \quad (137)$$

$$[V_{bad}]^2 = \int d\boldsymbol{k} \left[[f(\boldsymbol{k})]^2 m(\boldsymbol{k}) + 2[f(\boldsymbol{k})]^1 [m(\boldsymbol{k})]^1 + f(\boldsymbol{k})[m(\boldsymbol{k})]^2 \right] + H.c., \quad (138)$$

From these equations it follows that

$$[R, V] = [V]^1 = \int d\boldsymbol{k}_1 d\boldsymbol{k}_2 \left\{ \left[\hat{R}^{\boldsymbol{k}_2}, \hat{V}^{\boldsymbol{k}_1} \right] a(\boldsymbol{k}_2) a(\boldsymbol{k}_1) \right.$$
$$\left. - \left[\hat{R}^{\boldsymbol{k}_2 \dagger}, \hat{V}^{\boldsymbol{k}_1} \right] a^\dagger(\boldsymbol{k}_2) a(\boldsymbol{k}_1) + \hat{V}^{\boldsymbol{k}_1} \hat{R}^{\boldsymbol{k}_2 \dagger} \delta(\boldsymbol{k}_1 - \boldsymbol{k}_2) \right\} + H.c. \quad (139)$$

and

$$[R, [R, V]] = [V]^2 = \int d\boldsymbol{k}_1 d\boldsymbol{k}_2 d\boldsymbol{k}_3 \left\{ A_1(\boldsymbol{k}_1, \boldsymbol{k}_2, \boldsymbol{k}_3) a^\dagger(\boldsymbol{k}_2) a^\dagger(\boldsymbol{k}_1) a(\boldsymbol{k}_3) \right.$$
$$+ A_2(\boldsymbol{k}_1, \boldsymbol{k}_2, \boldsymbol{k}_3) a^\dagger(\boldsymbol{k}_2) a(\boldsymbol{k}_1) a(\boldsymbol{k}_3) + A_3(\boldsymbol{k}_1, \boldsymbol{k}_2, \boldsymbol{k}_3) a(\boldsymbol{k}_2) a(\boldsymbol{k}_1) a(\boldsymbol{k}_3)$$
$$\left. + A_4(\boldsymbol{k}_1, \boldsymbol{k}_2) a^\dagger(\boldsymbol{k}_2) \delta(\boldsymbol{k}_1 - \boldsymbol{k}_3) + A_5(\boldsymbol{k}_1, \boldsymbol{k}_2) a(\boldsymbol{k}_2) \delta(\boldsymbol{k}_1 - \boldsymbol{k}_3) \right\} + H.c., \quad (140)$$

where A_i are the following fermionic operators

$$A_1(\boldsymbol{k}_1, \boldsymbol{k}_2, \boldsymbol{k}_3) = \left[\hat{R}^{\boldsymbol{k}_3}, \left[\hat{R}^{\boldsymbol{k}_1}, \hat{V}^{\boldsymbol{k}_2} \right]^\dagger \right],$$

$$A_2(\boldsymbol{k}_1, \boldsymbol{k}_2, \boldsymbol{k}_3) = \left[\hat{R}^{\boldsymbol{k}_3}, \left[\hat{R}^{\boldsymbol{k}_1}, \hat{V}^{\boldsymbol{k}_2 \dagger} \right] + \left[\hat{R}^{\boldsymbol{k}_2}, \hat{V}^{\boldsymbol{k}_1 \dagger} \right]^\dagger \right],$$

$$A_3(\boldsymbol{k}_1, \boldsymbol{k}_2, \boldsymbol{k}_3) = \left[\hat{R}^{\boldsymbol{k}_3}, \left[\hat{R}^{\boldsymbol{k}_1}, \hat{V}^{\boldsymbol{k}_2} \right] \right],$$

$$A_4(\boldsymbol{k}_1, \boldsymbol{k}_2) = \hat{R}^{\boldsymbol{k}_1} \left\{ \left[\hat{R}^{\boldsymbol{k}_1}, \hat{V}^{\boldsymbol{k}_2} \right]^\dagger + \left[\hat{R}^{\boldsymbol{k}_2}, \hat{V}^{\boldsymbol{k}_1} \right]^\dagger \right\},$$

$$\begin{aligned} A_5(\boldsymbol{k}_1, \boldsymbol{k}_2) &= \hat{R}^{\boldsymbol{k}_1}\left[\hat{R}^{\boldsymbol{k}_1}, \hat{V}^{\boldsymbol{k}_2\dagger}\right]^{\dagger} + \left[\hat{R}^{\boldsymbol{k}_2}, \hat{R}^{\boldsymbol{k}_1}\right]\hat{V}^{\boldsymbol{k}_1\dagger} \\ &+2\hat{R}^{\boldsymbol{k}_1}\left[\hat{R}^{\boldsymbol{k}_2}, \hat{V}^{\boldsymbol{k}_1\dagger}\right] + \left[\hat{R}^{\boldsymbol{k}_2}, \hat{V}^{\boldsymbol{k}_1}\right]\hat{R}^{\boldsymbol{k}_1\dagger} \\ &+\hat{V}^{\boldsymbol{k}_1}\left[\hat{R}^{\boldsymbol{k}_2}, \hat{R}^{\boldsymbol{k}_1\dagger}\right]. \end{aligned} \tag{141}$$

For brevity, the subscript c has been omitted in the r.h.s. of Eqs. (139) and (140). In formula (141) we encounter ones and the same basic elements

$$\begin{aligned} \hat{V}^{\boldsymbol{k}} &= \int d\boldsymbol{p}d\boldsymbol{p}' \sum_{\mu,\mu',i,j} F_i^{\dagger}(\boldsymbol{p}', \mu')V_{ij}^{\boldsymbol{k}}(\boldsymbol{p}', \mu'; \boldsymbol{p}, \mu)F_j(\boldsymbol{p}, \mu), \\ \hat{R}^{\boldsymbol{k}} &= \int d\boldsymbol{p}d\boldsymbol{p}' \sum_{\mu,\mu',i,j} F_i^{\dagger}(\boldsymbol{p}', \mu')R_{ij}^{\boldsymbol{k}}(\boldsymbol{p}', \mu'; \boldsymbol{p}, \mu)F_j(\boldsymbol{p}, \mu), \end{aligned} \tag{142}$$

where $V_{ij}^{\boldsymbol{k}}$ and $R_{ij}^{\boldsymbol{k}}$ are a 2×2 c-number matrices

$$\begin{aligned} V_{i,j}^{\boldsymbol{k}}\left(\boldsymbol{p}', r'; \boldsymbol{p}, r\right) &= \frac{ig}{(2\pi)^{3/2}} \frac{m}{\sqrt{2\omega_{\boldsymbol{k}} E_{\boldsymbol{p}'} E_{\boldsymbol{p}}}} \delta\left(\boldsymbol{p} + \boldsymbol{k} - \boldsymbol{p}'\right) \\ &\times \begin{bmatrix} \bar{u}(\boldsymbol{p}', r')\gamma_5 u(\boldsymbol{p}, r) & \bar{u}(\boldsymbol{p}', r')\gamma_5 v(\boldsymbol{p}, r) \\ \bar{v}(\boldsymbol{p}', r')\gamma_5 u(\boldsymbol{p}, r) & \bar{v}(\boldsymbol{p}', r')\gamma_5 v(\boldsymbol{p}, r) \end{bmatrix}, \\ R_{i,j}^{\boldsymbol{k}}\left(\boldsymbol{p}', r'; \boldsymbol{p}, r\right) &= \frac{V_{i,j}^{\boldsymbol{k}}\left(\boldsymbol{p}', r'; \boldsymbol{p}, r\right)}{(-1)^{i-1}E'_{\boldsymbol{p}} - (-1)^{j-1}E_{\boldsymbol{p}} - \omega_k}, \end{aligned} \tag{143}$$

and the fermion column and row operators

$$F(\boldsymbol{p}, \mu) = \begin{pmatrix} b(\boldsymbol{p}, \mu) \\ d^{\dagger}(-\boldsymbol{p}, \mu) \end{pmatrix}, \quad F^{\dagger}(\boldsymbol{p}, \mu) = \left(b^{\dagger}(\boldsymbol{p}, \mu)\; d(-\boldsymbol{p}, \mu)\right). \tag{144}$$

Performing the normal ordering of the fermion operators in Eqs. (139) and (140), we get a simple recipe to select the $2 \longleftrightarrow 2$ and $2 \longleftrightarrow 3$ interaction operators of the g^2- and g^3-orders between the partially clothed pions, nucleon and antinucleons (in particular, $\pi N \to \pi N$, $NN \to NN$ and $NN \leftrightarrow \pi NN$). At the same time this algebraic technique enables to select the two-operator (one-body) contributions, which cancel the meson and fermion mass counterterms $M_{ren}(mes)$ and $M_{ren}(ferm)$ in the g^2-order (details are in [37] and [17]. In addition, there are three-operator (vertex-like) "radiative" corrections which together with the similar terms from the commutators $[R, M_{ren}(mes)]$ and $[R, M_{ren}(ferm)]$ cancel the "charge" counterterm V_{ren} in the g^3-order. The remaining bad terms must be removed via successive clothing UT's.

In its turn, the commutator $[V]^3 = [R,[R,[R,V]]] = [R,[V]^2]$, contains the g^4-order contribution to the $3N$-force operator $K_{NNN} \sim b^\dagger b^\dagger b^\dagger bbb$. The commutator $[V]^3$ being sandwiched $\langle 1'2'3'|[V]^3|123\rangle$ between the 3N-states is equivalent to $\langle 1'2'3'|[V]^3_{b^\dagger b^\dagger b^\dagger bbb}|123\rangle$ with

$$[V]^3_{b^\dagger b^\dagger b^\dagger bbb} = \int d\boldsymbol{k}_1 d\boldsymbol{k}_2 \{A_4(\boldsymbol{k}_1,\boldsymbol{k}_2)\hat{R}^{\boldsymbol{k}_2} + A_5(\boldsymbol{k}_1,\boldsymbol{k}_2)\hat{R}^{\boldsymbol{k}_2\dagger}\} + H.c. \,. \tag{145}$$

Just such matrix elements should be associated with the 3N-forces that enter the eigenvalue equation (127). Their non-relativistic reduction and application within the Faddeev formalism are under way. To conclude we would like to stress that this $3N$-quasipotential vanishes in the static limit so it has a relativistic origin.

3.5. Kharkov Potential in the Triton Binding Energy and p-d Scattering Observables Calculations

Recently, the Kharkov potential has been used for three-nucleon Faddeev calculations [12, 2] of the triton binding energy, momentum distributions of nucleons in the triton [3] and the elastic Nd scattering polarization observables at medium energies [13]. There 3N forces have been introduced via the Tucson-Melbourne model [7]. In this section we would like to show some of these results. These calculations gave the triton binding energy $E_T = -7.799$ MeV vs the value $E_T = -8.14\,(-8.150)$ MeV for Bonn (CD Bonn) potential compared to the experimental value -8.48 MeV. Perhaps, this discrepancy between the theory and experiment can be reduced after accounting for the $3N$-forces built up in the previous section. As seen in Fig. 7, the momentum distribution calculated with the UCT GS model has higher-momenta components compared to the rest ones.

The deuteron polarization observables (vector analyzing power iT_{11}, tensor analyzing power T_{20}, T_{21}, T_{22}) are shown in Fig. 8. As we can see, the three calculations for different nucleon-nucleon potentials reproduce fairly the measured asymmetries under consideration so it is difficult to

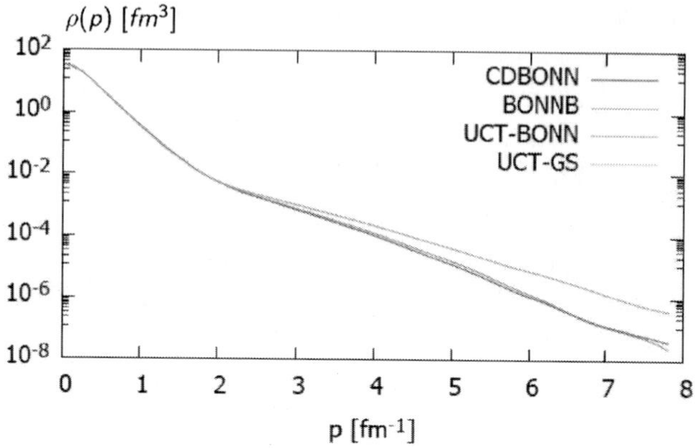

Figure 7. The momentum distribution of nucleons in the triton calculated with different NN potentials.

give preference to some of these models. At the same time Fig.9 demonstrates that the UCT GS + 3NF calculation gives the best description of this angular dependence removing to great extent the so-called Sagara-type discrepancy [26] in the vicinity of the cross section minimum. Of course, this result should not be overestimated because of an evident inconsistency between the $2N$-force and $3N$-force in question. In order to remove such an inconsistency we will extend our calculations with the $3N$-force interaction operator that has been discussed in the previous section.

4. QED in the Clothed-Particle Representation

Another field of our applications of the UCT method is QED. In the Coulomb gauge (used here) the interaction Hamiltonian of the spinor QED is given by (cf., for example, Eqs. (8.4.3) and (8.4.23) in [41])

$$V_{qed} = \int d\boldsymbol{x} V_{qed}(\boldsymbol{x}) = V_{qed}^{(1)} + V_{Coul}, \tag{146}$$

$$V_{qed}^{(1)} = \int d\boldsymbol{x} J^k(\boldsymbol{x}) A_k(\boldsymbol{x}),$$

with the electron-positron current density $J^\mu(\boldsymbol{x}) = e : \bar{\psi}(\boldsymbol{x})\gamma^\mu\psi(\boldsymbol{x}) :$, where the colon symbol denotes the normal ordering of the creation and

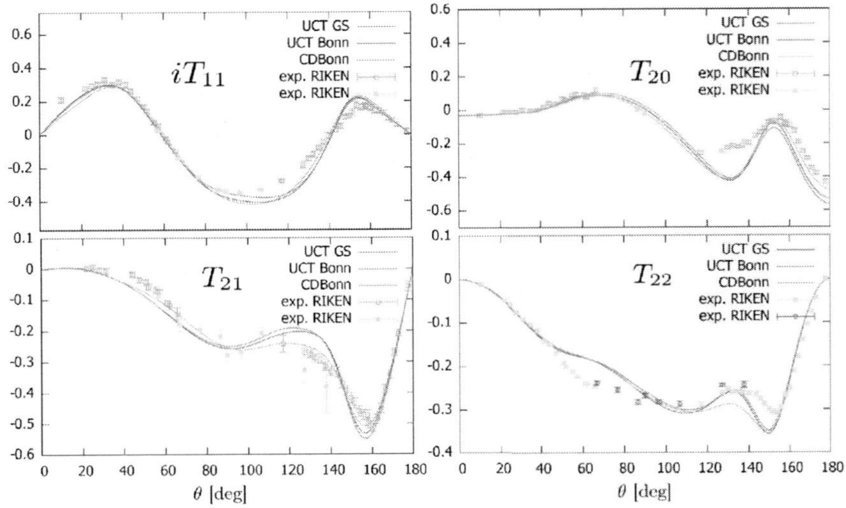

Figure 8. Polarization observables in the elastic Nd scattering: iT_{11}, T_{20}, T_{21} and T_{22} at E = 135 MeV/nucleon for different potentials. The experimental data are from [27, 28].

annihilation operators included. Its Coulomb part looks as

$$V_{Coul} = \frac{1}{2}\int d\boldsymbol{x}\int d\boldsymbol{y}\frac{J^0(\boldsymbol{x})J^0(\boldsymbol{y})}{4\pi|\boldsymbol{x}-\boldsymbol{y}|}e^{-\lambda|\boldsymbol{x}-\boldsymbol{y}|}. \tag{147}$$

Admittedly, the exponential factor with the parameter $\lambda > 0$ set to zero at the end of all calculations is introduced to deal with infrared divergences. Evidently, the interaction density $V_{qed}(x) = e^{iH_Ft}V_{qed}(\boldsymbol{x})e^{-iH_Ft}$ cannot be scalar, i.e., does not possesses the property (21). In this respect, for the Coulomb gauge (CG), where the photon field $A_\mu(\boldsymbol{x})$ is introduced in such a way to have $A_0(\boldsymbol{x}) \equiv 0$, we cannot use the so-called Belinfante ansatz to construct the boost generator $\boldsymbol{N}$, e.g., like in Eq. (27), therefore one has to seek other ways to provide the relativistic invariance (RI) in the Dirac sense (see, e.g., [35]).

In addition, we have the Fourier expansions for the photon field

$$A_\mu(\boldsymbol{x}) = \frac{1}{\sqrt{2(2\pi)^3}}\int\frac{d\boldsymbol{k}}{\omega_{\boldsymbol{k}}}\sum_\sigma\left(e_\mu(k\sigma)c(k\sigma)+e_\mu(k_-\sigma)c^\dagger(k_-\sigma)\right)e^{i\boldsymbol{k}\cdot\boldsymbol{x}}, \tag{148}$$

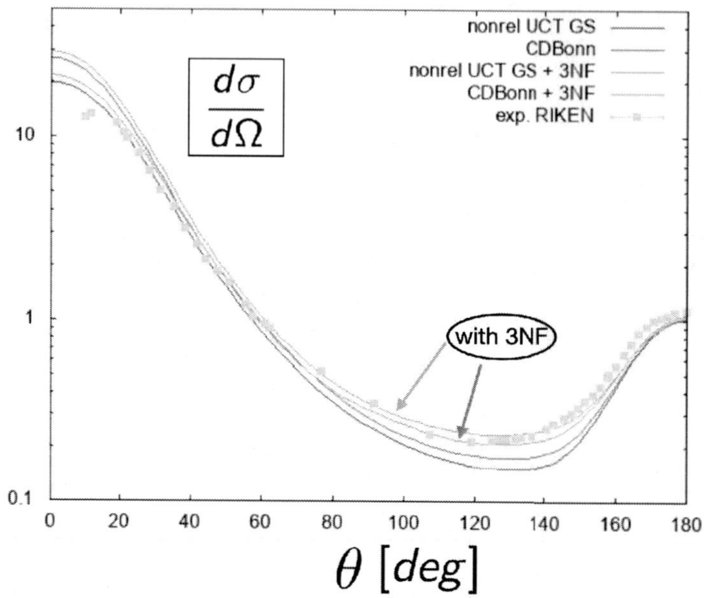

Figure 9. Differential cross section of pd elastic scattering as a function of the angle in c.m.s. for the proton incident energy 135 MeV calculated for different potentials with and without inclusion of the 3NF. The experimental data are from [27, 28].

where $p_- = (E_{\boldsymbol{p}}, -\boldsymbol{p})$, $k_- = (\omega_{\boldsymbol{k}}, -\boldsymbol{k})$, $\omega_{\boldsymbol{k}} = \sqrt{\boldsymbol{k}^2 + \lambda^2}$ the photon energy. The two independent 'polarization' vectors in Eq. (148) with the helicities $\sigma = \pm 1$, such that $\boldsymbol{k} \cdot \boldsymbol{e}(k\sigma) = 0$ and $e^0(k\sigma) = 0$, are normalized (see [41], Sect. 8.5), in terms of the timelike vector $n = (1, 0, 0, 0)$ so

$$\sum_\sigma e_\mu(k\sigma)e^*_\nu(k\sigma) = -g_{\mu\nu} + \omega_{\boldsymbol{k}} \frac{k_\mu n_\nu + k_\nu n_\mu}{\boldsymbol{k}^2} - \frac{k_\mu k_\nu}{\boldsymbol{k}^2} - \frac{k^2}{\boldsymbol{k}^2} n_\mu n_\nu \equiv P_{\mu\nu}(k)\,. \tag{149}$$

In the CPR the QED Hamiltonian can be written as

$$H_{qed} = K_F(\alpha_c) + K_I(\alpha_c), \tag{150}$$

where

$$K_F = \int \frac{d\boldsymbol{k}}{\omega_{\boldsymbol{k}}} \omega_{\boldsymbol{k}} c_c^\dagger(k) c_c(k) + \int \frac{d\boldsymbol{p}}{E_{\boldsymbol{p}}} E_{\boldsymbol{p}} \big[b_c^\dagger(p) b_c(p) + d_c^\dagger(p) d_c(p)\big], \tag{151}$$

while K_I contains of series of commutators responsible of all physical processes between clothed photons, electrons and positrons (cf. Eqs. (36)–(37)).

Following the clothing procedure exposed in Sec. 2.2 with the first clothing transformation $W^{(1)} = \exp(R^{(1)})$ we eliminate the primary interaction $V_{qed}^{(1)}$ since it consists of bad terms only. Retaining only the contributions in the e^2 order, the interaction operator can be written as (cf. Eq. (48))

$$\begin{aligned} K_I^{(2)}(\alpha_c) &= \frac{1}{2}\left[R_{qed}^{(1)}, V_{qed}^{(1)}\right] + V_{Coul} \\ &= K(e^-e^- \to e^-e^-) + K(e^+e^+ \to e^+e^+) \\ &+ K(e^-e^+ \to e^-e^+) + K(e^-e^+ \to \gamma\gamma) + K(\gamma\gamma \to e^-e^+) \\ &+ K(\gamma e^- \to \gamma e^-) + K(\gamma e^+ \to \gamma e^+)\,. \end{aligned} \tag{152}$$

Similarly the interacting vector mesons and nucleons primary non-scalar Coulomb interaction V_{Coul} cancels completely with contact terms that stem from the commutator $\frac{1}{2}\left[R_{qed}^{(1)}, V_{qed}^{(1)}\right]$. It is time to quote from [41] on p. 355, *viz.*, " ... the apparent violation of Lorentz invariance in the instantaneous Coulomb interaction is cancelled by another apparent violation of Lorentz invariance, ... " that arises since photon fields $A_D^{\mu}(x)$ do not transform as four-vectors, "and therefore have a non-covariant propagator." An important point is that in the CPR, unlike [41], such a cancellation takes place directly in the Hamiltonian. Such a distinct feature of the UCT method makes it useful in covariant calculations of the S-matrix either by solving the two-particle Lippmann-Schwinger equation (LSE) for the corresponding T-matrix or using the perturbation theory (not obligatorily addressing the Dyson-Feynman expansion). Of course, doing so one can find not only the S-matrix but the eigenstates of operator $K = K_F + K_I$ in the Fock subspace $R_F^{[2]}$ spanned onto the clothed-two-particle K_F eigenvectors.

4.1. Explicit Expressions for the Interaction Operators

In our previous works [4, 18] we have already presented the expressions for $K(e^-e^+ \to e^-e^+)$ and $K(e^-e^+ \to \gamma\gamma)$ when deriving the first correction to the positronium ground state energy and its decay to two photons. Now we will show the new e^2-order interactions between the clothed electrons, positrons and photons.

4.1.1. The Interaction Operator for Clothed Electrons

To get the operator $K(e^-e^- \to e^-e^-)$ one needs to separate out the $b_c^\dagger b_c^\dagger b_c b_c$-type terms from the Hamiltonian $K(\alpha_c)$

$$K(e^-e^- \to e^-e^-) \\ = \int \frac{d\boldsymbol{p}_1'}{E_{\boldsymbol{p}_1'}} \frac{d\boldsymbol{p}_2'}{E_{\boldsymbol{p}_2'}} \frac{d\boldsymbol{p}_1}{E_{\boldsymbol{p}_1}} \frac{d\boldsymbol{p}_2}{E_{\boldsymbol{p}_2}} V_{e^-e^-}(p_1', p_2'; p_1, p_2) b_c^\dagger(p_1') b_c^\dagger(p_2') b_c(p_1) b_c(p_2), \tag{153}$$

$$V_{e^-e^-}(p_1', p_2'; p_1, p_2) \\ = \frac{e^2 m^2}{(2\pi)^3} \delta(\boldsymbol{p}_2' + \boldsymbol{p}_1' - \boldsymbol{p}_2 - \boldsymbol{p}_1)[v_{e^-e^-}^{\text{Feynman-like}} + v_{e^-e^-}^{\text{off-energy-shell}}], \tag{154}$$

$$v_{e^-e^-}^{\text{Feynman-like}} = \frac{1}{2} \frac{\bar{u}(p_1')\gamma^\mu u(p_1) \bar{u}(p_2')\gamma_\mu u(p_2)}{(p_1' - p_1)^2 - \lambda^2}, \tag{155}$$

$$v_{e^-e^-}^{\text{off-energy-shell}} \\ = -\frac{1}{2} \frac{(p_1' + p_2' - p_1 - p_2)(p_1' - p_1)}{(\boldsymbol{p}_1' - \boldsymbol{p}_1)^2 + \lambda^2} \frac{\bar{u}(p_1')\gamma^0 u(p_1) \bar{u}(p_2')\gamma^0 u(p_2)}{(p_1' - p_1)^2 - \lambda^2}. \tag{156}$$

Here all momenta are defined on the mass-shell $p^2 = E_{\boldsymbol{p}}^2 - \boldsymbol{p}^2 = m^2$ (of course now with electron mass m).

The properly symmetrized interaction is given by the matrix element (quasipotential)

$$\bar{V}_{e^-e^-}(1', 2'; 1, 2) = \langle b_c^\dagger(1') b_c^\dagger(2') \Omega | K_{e^-e^- \to e^-e^-} | b_c^\dagger(1) b_c^\dagger(2) \Omega \rangle \\ = -V_{e^-e^-}(1', 2'; 1, 2) - V_{e^-e^-}(2', 1'; 2, 1) \\ + V_{e^-e^-}(2', 1'; 1, 2) + V_{e^-e^-}(1', 2'; 2, 1) \\ = \frac{e^2 m^2}{(2\pi)^3} \delta(\boldsymbol{p}_2' + \boldsymbol{p}_1' - \boldsymbol{p}_2 - \boldsymbol{p}_1) \left[\bar{v}_{e^-e^-}^{\text{Feynman-like}} + \bar{v}_{e^-e^-}^{\text{off-energy-shell}} \right], \tag{157}$$

with

$$\bar{v}_{e^-e^-}^{\text{Feynman-like}} = -\bar{u}(1')\gamma^\mu u(1)\frac{1}{2}\left\{\frac{1}{(p_1'-p_1)^2-\lambda^2} + \frac{1}{(p_2'-p_2)^2-\lambda^2}\right\}\bar{u}(2')\gamma_\mu u(2) - (1\leftrightarrow 2), \quad (158)$$

and

$$\bar{v}_{e^-e^-}^{\text{off-energy-shell}} = \frac{(p_1'+p_2'-p_1-p_2)^\mu}{(\boldsymbol{p}_1'-\boldsymbol{p}_1)^2+\lambda^2}\bar{u}(1')\gamma^0 u(1)\frac{1}{2}\left\{\frac{(p_1'-p_1)_\mu}{(p_1'-p_1)^2-\lambda^2} + \frac{(p_2'-p_2)_\mu}{(p_2'-p_2)^2-\lambda^2}\right\}\bar{u}(2')\gamma^0 u(2) - (1\leftrightarrow 2). \quad (159)$$

The non-covariant contribution (159) appears due to the interplay between the Coulomb interaction (147) and the non-covariant term from $\frac{1}{2}[R, V^{(1)}]_{b_c^\dagger b_c^\dagger b_c b_c}$. One should note, that on the energy shell for the electron-electron scattering, that is on the condition

$$E_{\boldsymbol{p}_1'} + E_{\boldsymbol{p}_2'} = E_{\boldsymbol{p}_1} + E_{\boldsymbol{p}_2} \quad (160)$$

this contribution vanishes. In this context, one should stress that on the energy shell only the Feynman-like contribution (158) remains and the expression in the curl brackets

$$\frac{1}{2}\left\{\frac{1}{(p_1'-p_1)^2-\lambda^2} + \frac{1}{(p_2'-p_2)^2-\lambda^2}\right\} \quad (161)$$

is converted into the genuine Feynman propagator which occurs when the S-operator in the e^2-order

$$S^{(2)}_{e^-e^-\to e^-e^-} = \int \frac{d\boldsymbol{p}_1'}{E_{\boldsymbol{p}_1'}}\frac{d\boldsymbol{p}_2'}{E_{\boldsymbol{p}_2'}}\frac{d\boldsymbol{p}_1}{E_{\boldsymbol{p}_1}}\frac{d\boldsymbol{p}_2}{E_{\boldsymbol{p}_2}} V^{\text{Feynman}}_{e^-e^-}(p_1', p_2'; p_1, p_2) b^\dagger(p_1') b^\dagger(p_2') b(p_1) b(p_2), \quad (162)$$

$$V^{\text{Feynman}}_{e^-e^-}(p_1', p_2'; p_1, p_2) = -i\frac{e_0^2 m_0^2}{(2\pi)^2}\delta(p_2'+p_1'-p_2-p_1) v^{\text{Feynman-like}}_{e^-e^-}(m\to m_0) \quad (163)$$

is sandwiched between the bare states $\langle b^\dagger(p_1') b^\dagger(p_2')\Omega_0 | S^{(2)} | b^\dagger(p_1) b^\dagger(p_2)\Omega_0\rangle$. The corresponding Feynman diagrams are displayed in Fig.10.

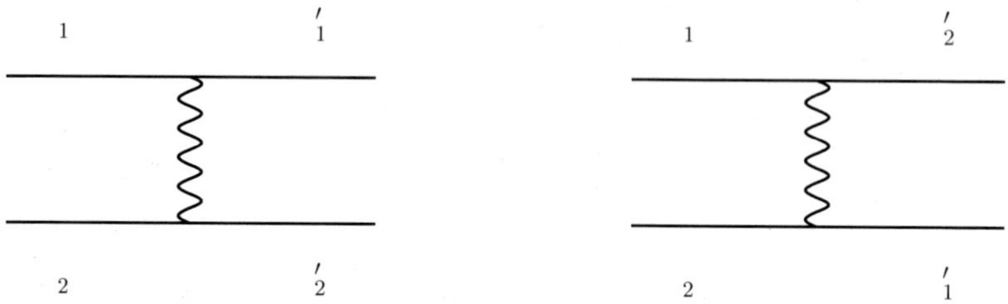

Figure 10. On-energy-shell contributions to the electron-electron interaction (the e^2 - order Feynman diagrams for the electron-electron scattering).

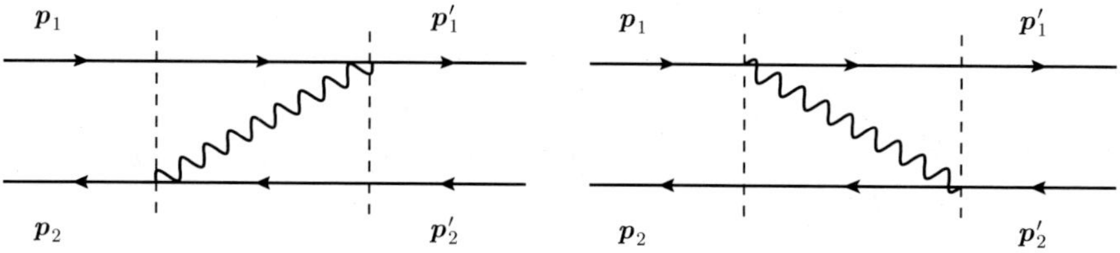

Figure 11. OFPT graphs for the scattering contribution (v_S) to the electron-positron scattering process.

It is important to emphasize that the interaction (157) is determined not only on the energy shell, but also beyond it, because it does not contain the factor $\delta(E_{\boldsymbol{p}_1}+E_{\boldsymbol{p}_2}-E_{\boldsymbol{p}_1'}-E_{\boldsymbol{p}_2'})$, that is a feature of the UCT method. The interaction obtained is nonlocal since the vertex factors and propagators in Eqs. (158) and (159) are dependent not only on the relative three-momenta involved but also on their total three-momentum. The interactions in [16] have the same property.

4.1.2. The Interaction Operator for Clothed Electron and Positron

Now we are interested in the $b_c^\dagger d_c^\dagger b_c d_c$-type terms

$$K(e^-e^+ \to e^-e^+)$$
$$= \int \frac{d\boldsymbol{p}_1'}{E_{\boldsymbol{p}_1'}} \frac{d\boldsymbol{p}_2'}{E_{\boldsymbol{p}_2'}} \frac{d\boldsymbol{p}_1}{E_{\boldsymbol{p}_1}} \frac{d\boldsymbol{p}_2}{E_{\boldsymbol{p}_2}} V_{e^-e^+}(p_1',p_2';p_1,p_2) b_c^\dagger(p_1') d_c^\dagger(p_2') b_c(p_1) d_c(p_2), \quad (164)$$

$$V_{e^-e^+}(p_1', p_2'; p_1, p_2) = \frac{e^2 m^2}{(2\pi)^3}\delta(\boldsymbol{p}_2' + \boldsymbol{p}_1' - \boldsymbol{p}_2 - \boldsymbol{p}_1)\left\{v_S(p_1', p_2'; p_1, p_2) + v_A(p_1', p_2'; p_1, p_2)\right\}, \tag{165}$$

$$v_{S/A} = v_{S/A}^{\text{Feynman-like}} + v_{S/A}^{\text{off-energy-shell}}, \tag{166}$$

with

$$v_S^{\text{Feynman-like}} = -\bar{u}(1')\gamma^\mu u(1)\frac{1}{2}\left\{\frac{1}{(p_1' - p_1)^2 - \lambda^2} + \frac{1}{(p_2' - p_2)^2 - \lambda^2}\right\}\bar{v}(2)\gamma_\mu v(2'),$$
$$v_S^{\text{off-energy-shell}} = \frac{(p_1' + p_2' - p_1 - p_2)^\mu}{(\boldsymbol{p}_1' - \boldsymbol{p}_1)^2 + \lambda^2}\bar{u}(1')\gamma^0 u(1)\frac{1}{2}\left\{\frac{(p_1' - p_1)_\mu}{(p_1' - p_1)^2 - \lambda^2} + \frac{(p_2' - p_2)_\mu}{(p_2' - p_2)^2 - \lambda^2}\right\}\bar{v}(2)\gamma^0 v(2'),$$

$$v_A^{\text{Feynman-like}} = \bar{u}(1')\gamma^\mu v(2')\frac{1}{2}\left\{\frac{1}{(p_1 + p_2)^2 - \lambda^2} + \frac{1}{(p_1' + p_2')^2 - \lambda^2}\right\}\bar{v}(2)\gamma_\mu u(1),$$
$$v_A^{\text{off-energy-shell}} = -\frac{(p_1' + p_2' - p_1 - p_2)^\mu}{(\boldsymbol{p}_1' + \boldsymbol{p}_2')^2 + \lambda^2}\bar{u}(1')\gamma^0 v(2')\frac{1}{2}\left\{\frac{(p_1' + p_2')_\mu}{(p_1' + p_2')^2 - \lambda^2} - \frac{(p_1 + p_2)_\mu}{(p_1 + p_2)^2 - \lambda^2}\right\}\bar{v}(2)\gamma^0 u(1). \tag{167}$$

Here in Eq. (165) we have introduced the decomposition into the so-called scattering and annihilation contributions v_S and v_A. Each of them has the structure (166) with Feynman-like and off-energy-shell terms. And, again, only the Feynman-like part survives on the energy-shell, i.e., on the condition $E_{\boldsymbol{p}_1'} + E_{\boldsymbol{p}_2'} = E_{\boldsymbol{p}_1} + E_{\boldsymbol{p}_2}$.

The scattering term v_S can be obtained from the direct parts of Eqs. (158)–(159) via substitution

$$\bar{u}(p_2') \to \bar{v}(p_2), \qquad u(p_2) \to v(p_2') \tag{168}$$

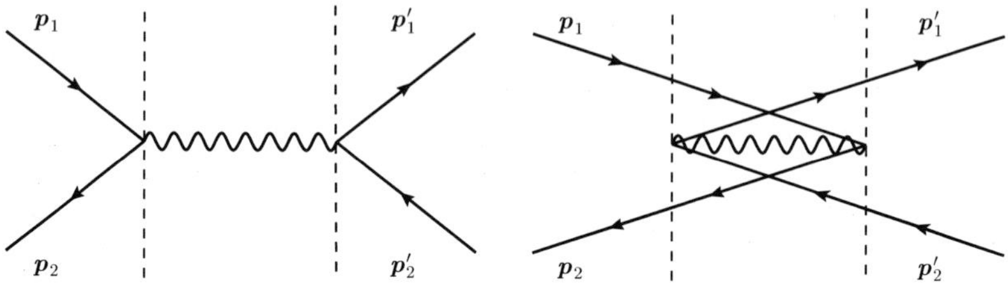

Figure 12. OFPT graphs for the annihilation contribution (v_A) to the electron-positron scattering process.

and the corresponding OFPT graphs are obtained from the is illustrated in Fig.11. Henceforth, the line directions in graphs are given with the sole scope to discriminate between the electron and positron states.

In the annihilation term the Feynman-like propagators arise from the non-covariant propagators in the following way

$$\begin{aligned}\frac{1}{\omega_{\boldsymbol{p}_1'+\boldsymbol{p}_2'}-E_{\boldsymbol{p}_1'}-E_{\boldsymbol{p}_2'}}+\frac{1}{\omega_{\boldsymbol{p}_1'+\boldsymbol{p}_2'}+E_{\boldsymbol{p}_1'}+E_{\boldsymbol{p}_2'}}&\rightarrow-\frac{1}{(p_1'+p_2')^2-\lambda^2},\\ \frac{1}{\omega_{\boldsymbol{p}_1+\boldsymbol{p}_2}-E_{\boldsymbol{p}_1}-E_{\boldsymbol{p}_2}}+\frac{1}{\omega_{\boldsymbol{p}_1+\boldsymbol{p}_2}+E_{\boldsymbol{p}_1}+E_{\boldsymbol{p}_2}}&\rightarrow-\frac{1}{(p_1+p_2)^2-\lambda^2}.\end{aligned}\tag{169}$$

The two non-covariant denominators

$$D_1^{-1}(E)_{E=E_{\boldsymbol{p}_1}+E_{\boldsymbol{p}_2}}\equiv(E-\omega_{\boldsymbol{p}_1+\boldsymbol{p}_2})|_{E=E_{\boldsymbol{p}_1}+E_{\boldsymbol{p}_2}}=-(\omega_{\boldsymbol{p}_1+\boldsymbol{p}_2}-E_{\boldsymbol{p}_1}-E_{\boldsymbol{p}_2}),$$
$$D_1^{-1}(E)_{E=E_{\boldsymbol{p}_1'}+E_{\boldsymbol{p}_2'}}\equiv(E-\omega_{\boldsymbol{p}_1'+\boldsymbol{p}_2'})|_{E=E_{\boldsymbol{p}_1'}+E_{\boldsymbol{p}_2'}}=-(\omega_{\boldsymbol{p}_1'+\boldsymbol{p}_2'}-E_{\boldsymbol{p}_1'}-E_{\boldsymbol{p}_2'})$$

can be associated with the left graph in Fig.12 In its turn, the right graph with five internal lines relates to the denominators

$$\begin{aligned}D_2^{-1}(E)_{E=E_{\boldsymbol{p}_1}+E_{\boldsymbol{p}_2}}&\equiv(E-E_{\boldsymbol{p}_1'}-E_{\boldsymbol{p}_2'}-E_{\boldsymbol{p}_1}-E_{\boldsymbol{p}_2}-\omega_{\boldsymbol{p}_1'+\boldsymbol{p}_2'})|_{E=E_{\boldsymbol{p}_1}+E_{\boldsymbol{p}_2}}\\&=-(\omega_{\boldsymbol{p}_1'+\boldsymbol{p}_2'}-E_{\boldsymbol{p}_1'}-E_{\boldsymbol{p}_2'}),\\ D_2^{-1}(E)_{E=E_{\boldsymbol{p}_1'}+E_{\boldsymbol{p}_2'}}&\equiv(E-E_{\boldsymbol{p}_1'}-E_{\boldsymbol{p}_2'}-E_{\boldsymbol{p}_1}-E_{\boldsymbol{p}_2}-\omega_{\boldsymbol{p}_1+\boldsymbol{p}_2})|_{E=E_{\boldsymbol{p}_1'}+E_{\boldsymbol{p}_2'}}\\&=-(\omega_{\boldsymbol{p}_1+\boldsymbol{p}_2}-E_{\boldsymbol{p}_1}-E_{\boldsymbol{p}_2}).\end{aligned}$$

The symmetrized interaction is

$$\begin{aligned}\bar{V}_{e^-e^+}(1',2';1,2)&=\langle b_c^\dagger(p_1')d_c^\dagger(p_2')\Omega|K_{e^-e^+\rightarrow e^-e^+}|b_c^\dagger(p_1)d_c^\dagger(p_2)\Omega\rangle\\&=-V_{e^-e^+}(p_1',p_2';p_1,p_2).\end{aligned}\tag{170}$$

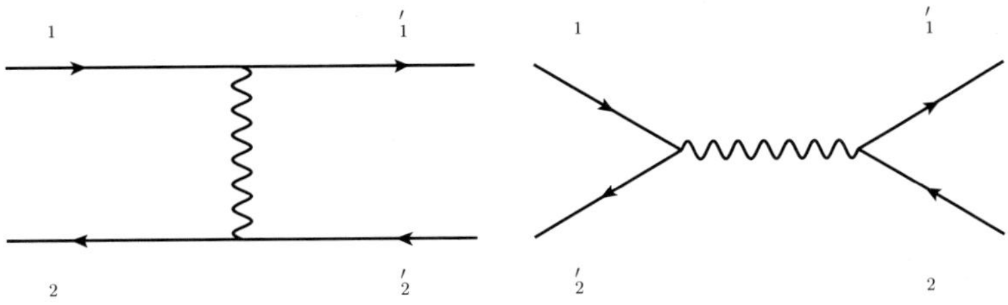

Figure 13. Feynman diagrams for the on-shell contribution to the electron-positron interaction.

As in case of the electron-electron interaction, the denominators in the curl brackets in the Feynman-like contributions are converted into the genuine Feynman propagators

$$\frac{1}{2}\left\{\frac{1}{(p_1'-p_1)^2-\lambda^2}+\frac{1}{(p_2'-p_2)^2-\lambda^2}\right\}\to\frac{1}{(p_1'-p_1)^2-\lambda^2}=\frac{1}{(p_2'-p_2)^2-\lambda^2},$$
$$\frac{1}{2}\left\{\frac{1}{(p_1+p_2)^2-\lambda^2}+\frac{1}{(p_1'+p_2')^2-\lambda^2}\right\}\to\frac{1}{(p_1'+p_2')^2-\lambda^2}=\frac{1}{(p_1+p_2)^2-\lambda^2},\tag{171}$$

displayed in Fig.13.

4.1.3. The Interaction Operator for e^-e^+-Pair Annihilation

The interaction operator corresponding to the annihilation of the clothed electron and positron in two photons is

$$K(e^-e^+\to\gamma\gamma)$$
$$=\int\frac{d\boldsymbol{k}_1}{\omega_{\boldsymbol{k}_1}}\frac{d\boldsymbol{k}_2}{\omega_{\boldsymbol{k}_2}}\frac{d\boldsymbol{p}_1}{E_{\boldsymbol{p}_1}}\frac{d\boldsymbol{p}_2}{E_{\boldsymbol{p}_2}}V_{e^-e^+\to\gamma\gamma}(k_2,k_1;p_2,p_1)c_c^\dagger(k_2)c_c^\dagger(k_1)b_c(p_2)d_c(p_1),\tag{172}$$

$$V_{e^-e^+\to\gamma\gamma}(k_2,k_1;p_2,p_1)$$
$$=\frac{e^2m}{2(2\pi)^3}\delta(\boldsymbol{p}_1+\boldsymbol{p}_2-\boldsymbol{k}_1-\boldsymbol{k}_2)[v_{e^-e^+\to\gamma\gamma}^{\text{Feynman-like}}+v_{e^-e^+\to\gamma\gamma}^{\text{off-energy-shell}}],\tag{173}$$

$$v^{\text{Feynman-like}}_{e^-e^+\to\gamma\gamma} = \frac{\bar{v}(p_1)\not{\epsilon}(k_1)\not{\epsilon}(k_2)u(p_2)}{\not{p}_1 - \not{k}_1 + m}, \tag{174}$$

$$v^{\text{off-energy-shell}}_{e^-e^+\to\gamma\gamma} = -\frac{1}{2}\left[\frac{\bar{v}(p_1)\not{\epsilon}(k_1)\not{\epsilon}(k_2)u(p_2)}{\not{p}_1 - \not{k}_1 + m} + \frac{\bar{v}(p_1)\not{\epsilon}(k_2)\not{\epsilon}(k_1)u(p_2)}{\not{p}_2 - \not{k}_1 - m}\right], \tag{175}$$

The corresponding matrix element (quasipotential) is determined as

$$\begin{aligned}
&\bar{V}_{e^-e^+\to\gamma\gamma}(k_2,k_1;p_2,p_1) = \\
&= \langle c_c^\dagger(k_2)c_c^\dagger(k_1)\Omega|K_{e^-e^+\to\gamma\gamma}|b_c^\dagger(p_2)d_c^\dagger(p_1)\Omega\rangle = \\
&\qquad = -V_{e^-e^+\to\gamma\gamma}(k_2,k_1;p_2,p_1) - V_{e^-e^+\to\gamma\gamma}(k_1,k_2;p_2,p_1) = \\
&\qquad = \frac{e^2 m}{2(2\pi)^3}\delta(\boldsymbol{p}_1 + \boldsymbol{p}_2 - \boldsymbol{k}_1 - \boldsymbol{k}_2)\bar{v}_{e^-e^+\to\gamma\gamma}(k_2,k_1;p_2,p_1),
\end{aligned} \tag{176}$$

$$\begin{aligned}
&\bar{v}_{e^-e^+\to\gamma\gamma}(k_2,k_1;p_2,p_1) = \\
&= \bar{v}(p_1)\not{\epsilon}(k_1)\frac{1}{2}\left\{\frac{1}{\not{p}_2 - \not{k}_2 - m} - \frac{1}{\not{p}_1 - \not{k}_1 + m}\right\}\not{\epsilon}(k_2)u(p_2) \\
&\qquad + \bar{v}(p_1)\not{\epsilon}(k_2)\frac{1}{2}\left\{\frac{1}{\not{p}_2 - \not{k}_1 - m} - \frac{1}{\not{p}_1 - \not{k}_2 + m}\right\}\not{\epsilon}(k_1)u(p_2).
\end{aligned} \tag{177}$$

The Feynman-like propagators in the curl brackets of this expression appear after summation of the corresponding non-covariant propagators which are illustrated with OFPT graphs in Fig.14. For example, by adding the contributions a) and b) in Fig.14 we get

$$\frac{m}{E_{\boldsymbol{k_1}-\boldsymbol{p_1}}}\left[\frac{P_+(q)}{E_{\boldsymbol{p}_1} - \omega_{\boldsymbol{k}_1} + E_{\boldsymbol{k_1}-\boldsymbol{p_1}}} + \frac{P_-(q_-)}{E_{\boldsymbol{p}_1} - \omega_{\boldsymbol{k}_1} - E_{\boldsymbol{k_1}-\boldsymbol{p_1}}}\right] = \frac{1}{\not{p}_1 - \not{k}_1 + m}, \tag{178}$$

with the four vector $q = (E_{\boldsymbol{k_1}-\boldsymbol{p_1}}, \boldsymbol{k}_1 - \boldsymbol{p}_1)$ and the projection operators on the fermion positive (negative)-energy states $P_\pm(q) = (\not{q} \pm m)/2m$. According to [16], we introduce "left" (s_1, u_1 and t_1) and "right" (s_2, u_2 and t_2) Mandelstam vectors

$$\begin{aligned}
s_1 = p_1 + k_1,\ s_2 = p_2 + k_2,\ u_1 = p_1 - k_2,& \\
u_2 = p_2 - k_1,\ t_1 = p_1 - k_1,\ t_2 = p_2 - k_2.&
\end{aligned} \tag{179}$$

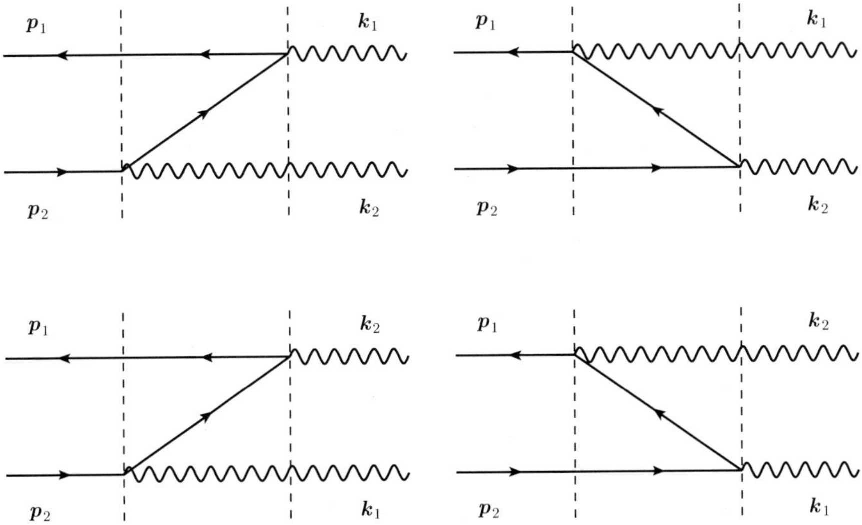

Figure 14. The direct (a,b) and exchange (c,d) OFPT diagrams for the interaction (176)

In these notations the expression (177) looks as

$$\bar{v}_{e^-e^+\gamma\gamma}(k_2, k_1; p_2, p_1) = \bar{v}(p_1)\not{e}(k_1)\frac{1}{2}\left\{\frac{1}{\not{t}_2 - m} - \frac{1}{\not{t}_1 + m}\right\}\not{e}(k_2)u(p_2)$$
$$+ \bar{v}(p_1)\not{e}(k_2)\frac{1}{2}\left\{\frac{1}{\not{u}_2 - m} - \frac{1}{\not{u}_1 + m}\right\}\not{e}(k_1)u(p_2). \quad (180)$$

Again on the energy shell $E_{\mathbf{p}_1} + E_{\mathbf{p}_2} = \omega_{\mathbf{k}_1} + \omega_{\mathbf{k}_2}$ the expressions in the curl brackets are transformed into the Feynman propagators

$$\begin{aligned} &\frac{1}{2}\left\{\frac{1}{\not{p}_2 - \not{k}_2 - m} - \frac{1}{\not{p}_1 - \not{k}_1 + m}\right\} \to \frac{1}{\not{p}_2 - \not{k}_2 - m}, \\ &\frac{1}{2}\left\{\frac{1}{\not{p}_2 - \not{k}_1 - m} - \frac{1}{\not{p}_1 - \not{k}_2 + m}\right\} \to \frac{1}{\not{p}_2 - \not{k}_1 - m}, \end{aligned} \quad (181)$$

that occur when evaluates the S-matrix for the e^+e^--pair annihilation. The corresponding Feynman diagrams are displayed in Fig.15. The operator (172) is not Hermitian and its Hermitian conjugate describe the e^+e^--pair

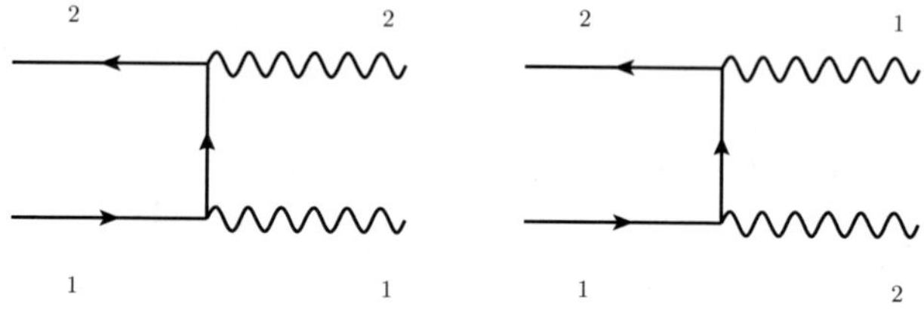

Figure 15. The Feynman diagrams of the e^2-order for the annihilation process.

production

$$K(\gamma\gamma \to e^- e^+) = \left[K(e^- e^+ \to \gamma\gamma)\right]^\dagger$$
$$= \int \frac{d\boldsymbol{k}_1}{\omega_{\boldsymbol{k}_1}} \frac{d\boldsymbol{k}_2}{\omega_{\boldsymbol{k}_2}} \frac{d\boldsymbol{p}_1}{E_{\boldsymbol{p}_1}} \frac{d\boldsymbol{p}_2}{E_{\boldsymbol{p}_2}} V_{\gamma\gamma \to e^- e^+}(p_2, p_1; k_2, k_1) b_c^\dagger(p_2) d_c^\dagger(p_1) c_c(k_2) c_c(k_1), \tag{182}$$

$$V_{\gamma\gamma \to e^- e^+}(p_2, p_1; k_2, k_1) = \frac{e^2 m}{2(2\pi)^3} \delta(\boldsymbol{p}_1 + \boldsymbol{p}_2 - \boldsymbol{k}_1 - \boldsymbol{k}_2)[\upsilon^{\text{Feynman-like}}_{\gamma\gamma \to e^- e^+} + \upsilon^{\text{off-energy-shell}}_{\gamma\gamma \to e^- e^+}], \tag{183}$$

$$\upsilon^{\text{Feynman-like}}_{\gamma\gamma \to e^- e^+} = -\frac{\bar{u}(p_2)\not{e}(k_1)\not{e}(k_2)v(p_1)}{\not{p}_1 - \not{k}_1 + m},$$
$$\upsilon^{\text{off-energy-shell}}_{\gamma\gamma \to e^- e^+} = \frac{1}{2}\left[\frac{\bar{u}(p_2)\not{e}(k_1)\not{e}(k_2)v(p_1)}{\not{p}_1 - \not{k}_1 + m} + \frac{\bar{u}(p_2)\not{e}(k_2)\not{e}(k_1)v(p_1)}{\not{p}_2 - \not{k}_1 - m}\right]. \tag{184}$$

To retain the Hermiticity of the total Hamiltonian the operators (172) and (182) should be considered jointly.

4.1.4. The Compton Effect in the Clothed Particle Representation

The Compton scattering is described by the operator with a structure $b_c^\dagger c_c^\dagger b_c c_c$

$$K(\gamma e^- \to \gamma e^-) =$$
$$= \int \frac{d\boldsymbol{k}_2}{\omega_{\boldsymbol{k}_2}} \frac{d\boldsymbol{p}_2}{E_{\boldsymbol{p}_2}} \frac{d\boldsymbol{k}_1}{\omega_{\boldsymbol{k}_1}} \frac{d\boldsymbol{p}_1}{E_{\boldsymbol{p}_1}} V_{\gamma e^-}(p_2, k_2; p_1, k_1) b_c^\dagger(p_2) c_c^\dagger(k_2) b_c(p_1) c_c(k_1), \tag{185}$$

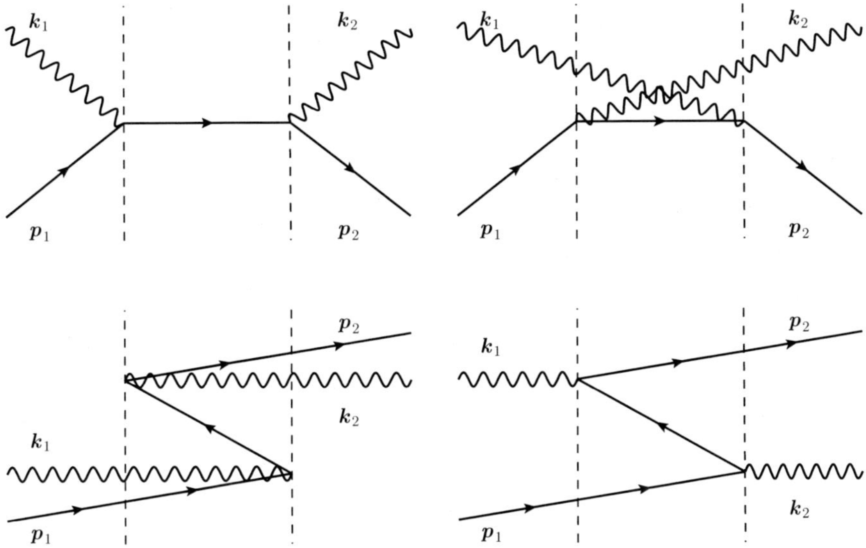

Figure 16. OFPT diagrams for different contributions to the interaction (188).

$$
\begin{aligned}
&V_{\gamma e^-}(p_2, k_2; p_1, k_1) = \\
&= \frac{e^2 m}{2(2\pi)^3}\delta(\boldsymbol{p}_1 + \boldsymbol{k}_1 - \boldsymbol{p}_2 - \boldsymbol{k}_2)[\upsilon_{\gamma e^-}^{\text{Feynman-like}} + \upsilon_{\gamma e^-}^{\text{off-energy-shell}}], \qquad (186)\\
&\upsilon_{\gamma e^-}^{\text{Feynman-like}} = \\
&= \bar{u}(p_2)\left\{\not{e}(k_2)\frac{1}{\not{p}_1 + \not{k}_1 - m}\not{e}(k_1) + \not{e}(k_1)\frac{1}{\not{p}_1 - \not{k}_2 - m}\not{e}(k_2)\right\}u(p_1),\\
&\upsilon_{\gamma e^-}^{\text{off-energy-shell}} = \\
&= \bar{u}(p_2)\not{e}(k_2)\frac{1}{2}\left\{\frac{1}{\not{k}_2 + \not{p}_2 - m} - \frac{1}{\not{p}_1 + \not{k}_1 - m}\right\}\not{e}(k_1)u(p_1)\\
&+ \bar{u}(p_2)\not{e}(k_1)\frac{1}{2}\left\{\frac{1}{\not{p}_2 - \not{k}_1 - m} - \frac{1}{\not{p}_1 - \not{k}_2 - m}\right\}\not{e}(k_2)u(p_1).
\end{aligned}
$$

(187)

The corresponding quasipotential is

Figure 17. Feynman diagrams for the Compton effect in the second order.

$$\bar{V}_{\gamma e^-}(p_2, k_2; p_1, k_1) = \\ = \langle c_c^\dagger(k_2) b_c^\dagger(p_2) | K_{\gamma e^- \to \gamma e^-} | c_c^\dagger(k_1) b_c^\dagger(p_1) \rangle = V_{\gamma e^-}(p_2, k_2; p_1, k_1). \tag{188}$$

To interpret the expression (188) we write an intermediate analytical result that leads to it

$$v_{\gamma e^-}(p_2, k_2; p_1, k_1) = \\ = \frac{1}{2} \bar{u}(p_2) \Big\{ \not{\epsilon}(k_2) P(p_2, k_2; p_1, k_1) \not{\epsilon}(k_1) + (k_2, p_2 \leftrightarrow k_1, p_1) \Big\} u(p_1), \tag{189}$$

where

$$P(p_2, k_2; p_1, k_1) = \\ = \frac{m}{E_{\boldsymbol{p}_1+\boldsymbol{k}_1}} \left\{ \frac{P_+(E_{\boldsymbol{p}_1+\boldsymbol{k}_1}, \boldsymbol{p}_1 + \boldsymbol{k}_1)}{E_{\boldsymbol{p}_1} + \omega_{\boldsymbol{k}_1} - E_{\boldsymbol{p}_1+\boldsymbol{k}_1}} + \frac{P_-(E_{\boldsymbol{p}_1+\boldsymbol{k}_1}, -\boldsymbol{p}_1 - \boldsymbol{k}_1)}{E_{\boldsymbol{p}_1} + \omega_{\boldsymbol{k}_1} + E_{\boldsymbol{p}_1+\boldsymbol{k}_1}} \right\} \\ + \frac{m}{E_{\boldsymbol{p}_2-\boldsymbol{k}_1}} \left\{ \frac{P_+(E_{\boldsymbol{p}_2-\boldsymbol{k}_1}, \boldsymbol{p}_2 - \boldsymbol{k}_1)}{E_{\boldsymbol{p}_2} - \omega_{\boldsymbol{k}_1} - E_{\boldsymbol{p}_2-\boldsymbol{k}_1}} + \frac{P_-(E_{\boldsymbol{p}_2-\boldsymbol{k}_1}, -\boldsymbol{p}_2 + \boldsymbol{k}_1)}{E_{\boldsymbol{p}_2} - \omega_{\boldsymbol{k}_1} + E_{\boldsymbol{p}_2-\boldsymbol{k}_1}} \right\}. \tag{190}$$

Each contribution to the r.h.s. of Eq. (190) can be represented by the graphs in Fig.16. The graphs a) and c) corresponds to the two terms in the first figure brackets:

$$D_a(E)|_{E=E_{\boldsymbol{p}_1}+\omega_{\boldsymbol{k}_1}} \equiv (E - E_{\boldsymbol{p}_1+\boldsymbol{k}_1})^{-1}_{E=E_{\boldsymbol{p}_1}+\omega_{\boldsymbol{k}_1}} = (E_{\boldsymbol{p}_1} + \omega_{\boldsymbol{k}_1} - E_{\boldsymbol{p}_1+\boldsymbol{k}_1})^{-1}, \\ D_c(E)|_{E=E_{\boldsymbol{p}_2}+\omega_{\boldsymbol{k}_2}} \equiv (E - E_{\boldsymbol{p}_1} - E_{\boldsymbol{p}_2} - \omega_{\boldsymbol{k}_1} - \omega_{\boldsymbol{k}_2} - E_{\boldsymbol{p}_1+\boldsymbol{k}_1})^{-1}_{E=E_{\boldsymbol{p}_2}+\omega_{\boldsymbol{k}_2}} \\ = -(E_{\boldsymbol{p}_1} + \omega_{\boldsymbol{k}_1} + E_{\boldsymbol{p}_1+\boldsymbol{k}_1})^{-1}, \tag{191}$$

while graphs b) and d) are associated with the two ones in the second figure brackets.

In terms of Mandelstam variables (179) the expression (189) is given by

$$v_{\gamma e^-}(p_2, k_2; p_1, k_1) = \bar{u}(p_2)\not{\epsilon}(k_2)\left\{\frac{1}{2}\frac{1}{\not{s}_1 - m} + \frac{1}{2}\frac{1}{\not{s}_2 - m}\right\}\not{\epsilon}(k_1)u(p_1)$$
$$+ \bar{u}(p_2)\not{\epsilon}(k_1)\left\{\frac{1}{2}\frac{1}{\not{u}_1 - m} + \frac{1}{2}\frac{1}{\not{u}_2 - m}\right\}\not{\epsilon}(k_2)u(p_1). \quad (192)$$

On the energy shell $E_{\mathbf{p}_1} + \omega_{\mathbf{k}_1} = E_{\mathbf{p}_2} + \omega_{\mathbf{k}_2}$, where there is no difference between left and right Mandelstam vectors, this equation reduces to

$$v_{e^-\gamma}(p_2, k_2; p_1, k_1)|_{E_{\mathbf{p}_1}+\omega_{\mathbf{k}_1}=E_{\mathbf{p}_2}+\omega_{\mathbf{k}_2}} =$$
$$= \frac{\bar{u}(p_2)\not{\epsilon}(k_2)\not{\epsilon}(k_1)u(p_1)}{\not{s} - m} + \frac{\bar{u}(p_2)\not{\epsilon}(k_1)\not{\epsilon}(k_2)u(p_1)}{\not{u} - m}, \quad (193)$$

which coincides up to numerical factor with the expressions, obtained with help of Feynman rules from the diagrams in Fig.17. One can replace Dirac spinors ($u \to v$) in the interactions (185) and (153) to obtain expressions for the corresponding processes with positrons.

Of course, all the results obtained here being taken on the energy-shell reproduce the well-known formulas for cross sections by Mller (the electron-electron scattering), KleinNishina (the Compton scattering) and Bhabha (the electron-positron scattering) (see, e.g., in [14]).

4.2. Positronium Properties in the Clothed Particle Representation

The interaction operator $K(e^-e^+ \to e^-e^+)$ from Sec. 4.1.2 between the clothed electron and positron has been used by us when solving the eigenvalue equation

$$\left[K_F^{(2)} + K(e^-e^+ \to e^-e^+)\right]|\Psi(\mathbf{P})\rangle = E|\Psi(\mathbf{P})\rangle, \quad (194)$$

for the vector $|\Psi\rangle$ from the electron-positron sector of the Fock space, spanned onto the two-clothed particle states $b_c^\dagger(p_1)d_c^\dagger(p_2)|\Omega\rangle$. Our purpose is to compare its solutions with the positronium spectrum found earlier (see [39] and refs. therein). Recall that the positronium ground state (g.s.) has two possible configurations with the total spin values $S = 0, 1$. The singlet (triplet) lowest-energy state with $S = 0$ $(S = 1)$ is known as the para-positronium (ortho-positronium). For this exposition, we will restrict our consideration of the para-positronium (p-Ps) configuration.

In the p-Ps rest system with the total momentum $\boldsymbol{P} = 0$ the eigenvalue equation for the wave function $\psi_{00}(\boldsymbol{p})$ of the para-positronium is reduced [18] to

$$2E_{\boldsymbol{p}}\psi_{00}(\boldsymbol{p}) + \int \frac{d\boldsymbol{p'}}{E_{\boldsymbol{p'}}E_{\boldsymbol{p}}}\bar{V}(p', p)\psi_{00}(\boldsymbol{p'}) = m_{\text{p-Ps}}\psi_{00}(\boldsymbol{p}), \tag{195}$$

where $m_{\text{p-Ps}} = m_{e^-} + m_{e^+} + \varepsilon_{\text{p-Ps}}$ the para-positronium mass and $\varepsilon_{\text{p-Ps}}$ its binding energy and $\bar{V}(p', p) = -V_{e^-e^+}(p', p'_-; p, p_-)$ with $p' = (E_{\boldsymbol{p'}}, \boldsymbol{p'})$, $p = (E_{\boldsymbol{p}}, \boldsymbol{p})$ and $p'_- = (E_{\boldsymbol{p'}}, -\boldsymbol{p'})$, $p_- = (E_{\boldsymbol{p}}, -\boldsymbol{p})$. In the non-relativistic limit $(E_{\boldsymbol{p}} = E_{\boldsymbol{p'}} = m)$ the eigenvalue equation is converted to the ordinary Schrödinger equation for the Coulomb potential in momentum space. Therefore we come to the well-known Coulomb problem with the g.s. energy $\varepsilon_{\text{g.s.}} \approx -6.8eV$. By considering the difference between $\bar{V}(p', p)$ and the Coulomb potential as a perturbation (it is not evident) and using the non-perturbative wave function of the ground state

$$\Psi_{00}(\boldsymbol{p}) = \frac{2}{\pi}\frac{\sqrt{2a^3}}{(1 + a^2\boldsymbol{p}^2)^2}, \tag{196}$$

from Appendix C in [39] we have obtained the energy shift

$$\Delta\varepsilon = -4.7325 \cdot 10^{-4}\ eV.$$

This value surprisingly coincides with those estimations given in [39] (see formula (1.1) therein). In order to verify such a coincidence beyond the perturbation theory, one needs to solve Eq. (195) numerically. In this context, similarly the deuteron in Sec. 3.2, the partial eigenvalue equation for the para-positronium WFs that belong to the total angular momentum J takes the form

$$2E_{\boldsymbol{p}}\Psi^J(p) + \int_0^\infty \frac{p'^2 dp'}{E_{\boldsymbol{p'}}E_{\boldsymbol{p}}}\Psi^J(p')\bar{V}^J(p, p') = m\text{p-Ps}\Psi^J(p). \tag{197}$$

Here $\bar{V}^J(p, p')$ is the partial electron-positron quasipotential derived in the momentum representation from the new e^-e^+-interaction operator. In turn, we have

$$\bar{V}^J(p,p') = \frac{e^2}{4\pi^2}\left[\bar{v}^J_{\text{Feynman-like}} + \bar{v}^J_{\text{off-shell}}\right], \tag{198}$$

with

$$\begin{aligned} \bar{v}^J_{\text{Feynman-like}} &= -\frac{2E_{\boldsymbol{p}'}E_{\boldsymbol{p}} - m^2}{pp'}Q_J(z_2), \\ \bar{v}^J_{\text{off-shell}} &= -\frac{(E_{\boldsymbol{p}'} + E_{\boldsymbol{p}})^2}{2pp'}Q_J(z_1) + \frac{2E_{\boldsymbol{p}'}E_{\boldsymbol{p}}}{pp'}Q_J(z_2), \end{aligned} \tag{199}$$

where $Q_J(z)$ the Legendre function of the second kind and

$$z_1 = \frac{p'^2 + p^2 + \lambda^2}{2pp'}, \quad z_2 = \frac{E_{\boldsymbol{p}'}E_{\boldsymbol{p}} - m^2 + \frac{1}{2}\lambda^2}{pp'}.$$

Such a separation implies that only the Feynman-like part survives on the energy shell, where $E_{\boldsymbol{p}'} = \sqrt{p'^2 + m^2} = E_{\boldsymbol{p}} = \sqrt{p^2 + m^2}$. The task of solving the eigenvalue equation and obtaining the corresponding positronium states in the CPR is underway.

4.2.1. Positronium Decay Rates

We have started with Eq. (9.337) in [14] that determines the decay rates of interest

$$\Gamma = \sum_{\sigma_1\,\sigma_2}\int\frac{d\boldsymbol{k}_1}{k_1^0}\int\frac{d\boldsymbol{k}_2}{k_2^0}\,\pi\delta(k_1^0 + k_2^0 - E_{\text{Ps}})\delta(\boldsymbol{k}_1 + \boldsymbol{k}_2 - \boldsymbol{P})|T_{fi}|^2, \tag{200}$$

where the quantity $T_{fi} = \langle\Omega|c_c(k_1\sigma_1)c_c(k_2\sigma_2)T|p\text{-}Ps\rangle$ denotes the T-matrix element for transition from the initial p-Ps ground state to the final state of the two photons with the momenta $k_1 = (k_1^0, \boldsymbol{k}_1)$ and $k_2 = (k_2^0, \boldsymbol{k}_2)$ and their polarizations σ_1, σ_2. In this connection, we note an equivalence theorem proved in [31], that allows us to use a recipe for calculating the S-matrix (T-matrix) in the CPR.

In the rest frame of positronium ($\boldsymbol{P} = 0$) one can do integration with both δ-functions

$$\Gamma = \frac{\pi}{2}\sum_{\sigma_1\,\sigma_2}\int d\hat{\boldsymbol{k}}_1|T_{fi}|^2, \tag{201}$$

where $\hat{\boldsymbol{k}}_1$ is the unit vector along vector $\boldsymbol{k}_1$.

The positronium state-vector with spin S can be represented as

$$|Ps(S)\rangle = \sum_{M_S} \int \frac{d\boldsymbol{p}}{p^0} \Psi_{SM_S}(\boldsymbol{p}) |p\, SM_S\rangle \,, \tag{202}$$

where $|p\, SM_S\rangle = \sum\limits_{\mu_1\mu_2} (\frac{1}{2}\mu_1 \frac{1}{2}\mu_2 | SM_S) b_c^\dagger(p_-\mu_2) d_c^\dagger(p\mu_1)|\Omega\rangle$ the electron-positron state-vector in the total spin representation and $p_- = (E_{\boldsymbol{p}}, -\boldsymbol{p})$. In the Born approximation where $T \approx K_{e^-e^+\to\gamma\gamma}$ one has

$$\Gamma = \frac{\pi}{2} \sum_{\sigma_1\,\sigma_2} \int d\hat{\boldsymbol{k}} |\langle k\sigma_1\sigma_2 | K_{e^-e^+\to\gamma\gamma} | \text{p-Ps}\rangle|^2 \,,$$

$$\langle k\sigma_1\sigma_2 | K_{e^-e^+\to\gamma\gamma} | \text{p-Ps}\rangle = \int \frac{d\boldsymbol{p}}{E_{\boldsymbol{p}}} \Psi_{00}(\boldsymbol{p}) \langle k\sigma_1\sigma_2 | K_{e^-e^+\to\gamma\gamma} | p\, 0\, 0\rangle \,, \tag{203}$$

where $\langle k\sigma_1\sigma_2| \equiv \langle\Omega| c_c(k\sigma_1) c_c(k_-\sigma_2)$. Here the p-Ps g.s. is approximated with the function (196).

The interaction (172) in an arbitrary frame has the form

$$\langle\Omega| c_c(k_1\sigma_1) c_c(k_2\sigma_2) K_{e^-e^+\to\gamma\gamma} b_c^\dagger(p_2\mu_2) d_c^\dagger(p_1\mu_1)|\Omega\rangle = \frac{\alpha m}{4\pi^2} \bar{v}_{e^-e^+\gamma\gamma} \,, \tag{204}$$

$$\bar{v}_{e^-e^+\gamma\gamma} = \bar{v}(p_1\mu_1) \not{e}(k_1\sigma_1) \frac{1}{2} \left\{ \frac{1}{\not{p}_2 - \not{k}_2 - m} - \frac{1}{\not{p}_1 - \not{k}_1 + m} \right\} \not{e}(k_2\sigma_2) u(p_2\mu_2) + \\ + (k_1, \sigma_1 \leftrightarrow k_2, \sigma_2) \,, \tag{205}$$

whence in the static limit ($\boldsymbol{p}_{1,2} = 0$) we arrive to the well-known Pirenne-Wheeler result with $\Gamma = \frac{1}{2}\alpha^5 m \approx 8.0325 \cdot 10^9$ sec^{-1} [42, 25]. After this by using the CG we get

$$\Gamma = \frac{128}{\pi^2} m\alpha^7 \mathrm{I}(\alpha)^2 \tag{206}$$

with the integral

$$\mathrm{I}(\alpha) = \frac{1}{8} \int\limits_0^\infty du \, \frac{u \sinh u}{(\alpha^2/4 + \sinh^2 u)^2} = 26.7535 \,. \tag{207}$$

Now we get the decay rate of para-positronium into two photons $\Gamma = 7.9411 \cdot 10^9$ sec^{-1} . The experimental result [1] for this value $7.9909 \pm 0.0017 \cdot 10^9$ sec^{-1}.

In our report [33] we have discussed a possible application of the UCT method in the quantum chromodynamics (QCD) by drawing some parallels between the total QCD Hamiltonian and the QED Hamiltonian in the Coulomb gauge. In this respect, we foresee a new research work with good prospects.

4.3. Mass Renormalization in Mesodynamics and Quantum Electrodynamics

In the field models under the consideration the operator $M_{ren} = M_{ren,mes} + M_{ren,\ ferm}$ consists of the meson counterterms

$$M_{ren,mes} = \frac{\mu_0^2 - \mu^2}{4} \int \frac{d\boldsymbol{k}}{\omega_{\boldsymbol{k}}} \left[a^\dagger(\boldsymbol{k})\, a(\boldsymbol{k}) + a(\boldsymbol{k})\, a(-\boldsymbol{k}) + H.c.\right] \tag{208}$$

and the fermion ones

$$M_{ren,\ ferm} = m(m_0 - m) \int \frac{d\boldsymbol{p}}{E_{\boldsymbol{p}}} \sum_{\mu\mu',ij} F_i^\dagger(\boldsymbol{p}\mu')\, M_{ij}(\boldsymbol{p}\mu';\boldsymbol{p}\mu)\, F_j(\boldsymbol{p}\mu), \tag{209}$$

with $M_{ij}(\boldsymbol{p}\mu';\boldsymbol{p}\mu) = \bar{U}_i(\boldsymbol{p}\mu')\, U_j(\boldsymbol{p}\mu)$, where $\bar{U}_i(\boldsymbol{p}\mu')$ $(i = 1, 2)$ is the element of the row $[\bar{u}(\boldsymbol{p}\mu), \bar{v}(-\boldsymbol{p}\mu)]$. Here and henceforth in this Sec. we omit the subscript c for the clothed operators.

In both cases of couplings (13)-(15) in mesodynamics and couplings (146) in QED we consider only terms of the second order in coupling constants. Doing so we get from Eq. (37)

$$K_I^{(2)}(\alpha_c) = M_{\text{ren}}^{(2)} + V^{(2)} + \frac{1}{2}\left[R^{(1)}, V^{(1)}\right]. \tag{210}$$

Moreover, the commutator $\left[R^{(1)}, V^{(1)}\right]$ also contains the meson and fermion two-operator terms $a^\dagger a$, aa, $b^\dagger b$, $b^\dagger d^\dagger$ and so on, whose structure repeats the structure of the Eqs. (208) and (209). Not all of them are bad (for instance, $a^\dagger a$). It is required that the “diagonal” (particle-conserving number) species of the $a^\dagger a$-, $b^\dagger b$-, and $d^\dagger d$- types cancel the corresponding contributions to the mass counterterms $M_{ren,mes}(\alpha_c)$ and

$M_{ren,\,ferm}\,(\alpha_c)$. Note that it is sufficient to evaluate the mass shifts $m - m_0$ and $\mu^2 - \mu_0^2$ in the g^2-order; since the same operator structure will appear in higher orders in coupling constants we can extend this requirement to determine these mass shifts order by order.

The first results in this direction have been obtained in [37], [17] for interacting pions and nucleons with the PS coupling (13). For example, according to [37], the PS meson mass shift in the g^2 -order is equal to

$$\delta\mu^2 \equiv \mu_0^2 - \mu^2 = \frac{2g_{ps}^2}{(2\pi)^3} \int \frac{d\boldsymbol{p}}{E_{\boldsymbol{p}}} \left\{ 1 + \frac{\mu^4}{4\,(pk)^2 - \mu^4} \right\}. \tag{211}$$

The nucleon mass shift due to PS meson exchange evaluated in [17] in the same order can be written as

$$\begin{aligned} \delta m_N &\equiv m_0 - m = \frac{g_{ps}^2}{4m(2\pi)^3} \left[I_1(p) + I_2(p) \right], \\ I_1(p) &= \int \frac{d\boldsymbol{k}}{\omega_{\boldsymbol{k}}} pk \left\{ \frac{1}{\mu^2 - 2pk} - \frac{1}{\mu^2 + 2pk} \right\}, \\ I_2(p) &= \int \frac{d\boldsymbol{q}}{E_{\boldsymbol{q}}} \left\{ \frac{m^2 - pq}{2\left[m^2 - pq\right] - \mu^2} + \frac{m^2 + pq}{2\left[m^2 + pq\right] - \mu^2} \right\}. \end{aligned} \tag{212}$$

We observe that, being expressed through the explicitly covariant integrals in the r.h.s of Eqs. (211) and (212), these quantities do not depend on the particle momenta. It turns out [17] that these integrals coincide with the corresponding one-loop Feynman integrals.

Further, we would like to show the way to calculate the fermion mass shifts due to vector boson exchange. As mentioned above, all particle-conserving mass renormalization counterterms in $M_{\text{ren}}^{(2)} = M_{\text{boson}}^{(2)} + M_{\text{fermion}}^{(2)}$ are cancelled by the corresponding components of the commutator $\frac{1}{2}\left[R_{\text{v}}^{(1)}, V_{\text{v}}^{(1)}\right]$ and the operator $V_{\text{v}}^{(2)}$. For fermions, it gives us the equation

$$M_{\text{ren}}^{(2)}{}_{b^\dagger b} + V_{\text{v}}^{(2)}{}_{b^\dagger b} + \frac{1}{2}\left[R_{\text{v}}^{(1)}, V_{\text{v}}^{(1)}\right]_{b^\dagger b} = 0\,. \tag{213}$$

Subscript $b^\dagger b$ means that after normal ordering, we retain only one-fermion terms. Note that the equation is the same for one-antifermion terms ($d^\dagger d$), which means the same results for antifermion mass shifts. In this context, let us write down all the one-fermion terms included in the equation (213)

$$M^{(2)}_{\mathrm{ren}\ b^\dagger b} = m\delta m \int \frac{d\boldsymbol{p}}{E^2_{\boldsymbol{p}}} \sum_{\rho} b^\dagger(p\rho) b(p\rho)\,, \tag{214}$$

$$\begin{aligned}
\frac{1}{2}\big[R^{(1)}_{\mathrm{v}}, V^{(1)}_{\mathrm{v}}\big]_{b^\dagger b} &= \frac{m}{8(2\pi)^3} \int \frac{d\boldsymbol{p}}{E^2_{\boldsymbol{p}}} \frac{d\boldsymbol{q}}{E_{\boldsymbol{q}}\omega_{\boldsymbol{p}-\boldsymbol{q}}} \sum_{\mu\,\rho} b^\dagger(p\rho) b(p\rho) \\
&\times \bar{u}(p\mu)\Bigg\{ \frac{P^{\alpha\beta}_{\mathrm{v}}(k)}{E_{\boldsymbol{p}} - \omega_{\boldsymbol{p}-\boldsymbol{q}} - E_{\boldsymbol{q}}} \left[g_{\mathrm{v}}\,\gamma_\alpha - \frac{f_{\mathrm{v}}}{2m} i k^\xi \sigma_{\xi\alpha} \right] (\not{q} + m) \left[g_{\mathrm{v}}\,\gamma_\beta - \frac{f_{\mathrm{v}}}{2m} i k^\eta \sigma_{\beta\eta} \right] \\
&+ \frac{P^{\alpha\beta}_{\mathrm{v}}(k_-)}{E_{\boldsymbol{p}} + \omega_{\boldsymbol{p}-\boldsymbol{q}} + E_{\boldsymbol{q}}} \left[g_{\mathrm{v}}\,\gamma_\alpha - \frac{f_{\mathrm{v}}}{2m} i k^\xi_- \sigma_{\alpha\xi} \right] (\not{q}_- - m) \left[g_{\mathrm{v}}\,\gamma_\beta - \frac{f_{\mathrm{v}}}{2m} i k^\eta_- \sigma_{\eta\beta} \right] \Bigg\} u(p\mu),
\end{aligned} \tag{215}$$

$$\begin{aligned}
V^{(2)}_{\mathrm{v}\ b^\dagger b} &= \frac{m}{8(2\pi)^3} \int \frac{d\boldsymbol{p}}{E^2_{\boldsymbol{p}}} \frac{d\boldsymbol{q}}{E_{\boldsymbol{q}}\omega_{\boldsymbol{p}-\boldsymbol{q}}} \sum_{\mu\,\rho} b^\dagger(p\rho) b(p\rho) \\
&\times \bar{u}(p\mu)\Bigg\{ g^2_{\mathrm{v}} \frac{\omega_{\boldsymbol{p}-\boldsymbol{q}}}{m^2_b} \Big[\gamma_0(\not{q} + m)\gamma_0 - \gamma_0(\not{q}_- - m)\gamma_0 \Big] \\
&- f^2_{\mathrm{v}} \frac{\omega_{\boldsymbol{p}-\boldsymbol{q}}}{4m^2} \Big[\gamma_0\boldsymbol{\gamma}(\not{q} + m)\gamma_0\boldsymbol{\gamma} - \gamma_0\boldsymbol{\gamma}(\not{q}_- - m)\gamma_0\boldsymbol{\gamma} \Big] \Bigg\} u(p\mu)\,.
\end{aligned} \tag{216}$$

The r.h.s of Eq. (216) embodies only the so-called contact terms in which the propagators of intermediate particles are replaced by squares of their masses. Keeping in mind the definition (24) of the projection operator $P^{\alpha\beta}_{\mathrm{v}}(k)$

$$\begin{aligned}
\frac{1}{2}\big[R^{(1)}_{\mathrm{v}}, V^{(1)}_{\mathrm{v}}\big]_{b^\dagger b} &= -\frac{m}{8(2\pi)^3} \int \frac{d\boldsymbol{p}}{E^2_{\boldsymbol{p}}} \frac{d\boldsymbol{q}}{E_{\boldsymbol{q}}\omega_{\boldsymbol{p}-\boldsymbol{q}}} \sum_{\mu\,\rho} b^\dagger(p\rho) b(p\rho) \\
&\times \bar{u}(p\mu)\Bigg\{ \frac{1}{E_{\boldsymbol{p}} - \omega_{\boldsymbol{p}-\boldsymbol{q}} - E_{\boldsymbol{q}}} \left[\gamma_\alpha(g_{\mathrm{v}} + f_{\mathrm{v}}) - \frac{f_{\mathrm{v}}}{2m}(p+q)_\alpha \right] (\not{q} + m) \\
&\qquad \times \left[\gamma^\alpha(g_{\mathrm{v}} + f_{\mathrm{v}}) - \frac{f_{\mathrm{v}}}{2m}(p+q)^\alpha \right] \\
&+ \frac{1}{E_{\boldsymbol{p}} + \omega_{\boldsymbol{p}-\boldsymbol{q}} + E_{\boldsymbol{q}}} \left[\gamma_\alpha(g_{\mathrm{v}} + f_{\mathrm{v}}) - \frac{f_{\mathrm{v}}}{2m}(p-q_-)_\alpha \right] (\not{q}_- - m)
\end{aligned}$$

$$\times \left[\gamma^\alpha(g_\text{v}+f_\text{v}) - \frac{f_\text{v}}{2m}(p-q_-)^\alpha\right] + f_\text{v}\frac{E_{\boldsymbol{q}}}{m^2}[6m(g_\text{v}+f_\text{v}) + f_\text{v}(\boldsymbol{p}+\boldsymbol{q})\boldsymbol{\gamma}]$$

$$+ g_\text{v}^2\frac{\omega_{\boldsymbol{p}-\boldsymbol{q}}}{m_b^2}\left[\gamma_0(\not{q}+m)\gamma_0 - \gamma_0(\not{q}_- - m)\gamma_0\right]$$

$$- f_\text{v}^2\frac{\omega_{\boldsymbol{p}-\boldsymbol{q}}}{4m^2}\left[\gamma_0\boldsymbol{\gamma}(\not{q}+m)\gamma_0\boldsymbol{\gamma} - \gamma_0\boldsymbol{\gamma}(\not{q}_- - m)\gamma_0\boldsymbol{\gamma}\right]\Bigg\}u(p\mu)\,, \quad (217)$$

we can see that the expression (216) cancels completely the last two contact terms in Eq. (217).

In the case of QED: $g = e$, $f = 0$, $m_b = \lambda$ and

$$V_\text{v}^{(2)}{}_{b^\dagger b} = \frac{e^2 m}{8(2\pi)^3}\int \frac{d\boldsymbol{p}}{E_{\boldsymbol{p}}^2}\frac{d\boldsymbol{q}}{E_{\boldsymbol{q}}\omega_{\boldsymbol{p}-\boldsymbol{q}}}\sum_{\mu\,\rho} b^\dagger(p\rho)b(p\rho)\frac{\omega_{\boldsymbol{p}-\boldsymbol{q}}}{(\boldsymbol{p}-\boldsymbol{q})^2+\lambda^2}$$

$$\times\, \bar{u}(p\mu)\left[\gamma_0(\not{q}+m)\gamma_0 - \gamma_0(\not{q}_- - m)\gamma_0\right]u(p\mu) \equiv V_{\text{Coul}\,b^\dagger b}\,. \quad (218)$$

By introducing the photon polarization tensor $P_{\alpha\beta}(k)$ (149) we arrive to

$$\frac{1}{2}[R_{qed}^{(1)}, V_{qed}^{(1)}]_{b^\dagger b} = -\frac{e^2 m}{8(2\pi)^3}\int \frac{d\boldsymbol{p}}{E_{\boldsymbol{p}}^2}\frac{d\boldsymbol{q}}{E_{\boldsymbol{q}}\omega_{\boldsymbol{p}-\boldsymbol{q}}}\sum_{\mu\,\rho} b^\dagger(p\rho)b(p\rho)$$

$$\times\, \bar{u}(p\mu)\left\{\frac{1}{E_{\boldsymbol{p}}-\omega_{\boldsymbol{p}-\boldsymbol{q}}-E_{\boldsymbol{q}}}\gamma_\alpha(\not{q}+m)\gamma^\alpha + \frac{1}{E_{\boldsymbol{p}}+\omega_{\boldsymbol{p}-\boldsymbol{q}}+E_{\boldsymbol{q}}}\gamma_\alpha(\not{q}_- - m)\gamma^\alpha\right.$$

$$\left. + \frac{\omega_{\boldsymbol{p}-\boldsymbol{q}}}{(\boldsymbol{p}-\boldsymbol{q})^2}\left[\gamma_0(\not{q}+m)\gamma_0 - \gamma_0(\not{q}_- - m)\gamma_0\right]\right\}u(p\mu)\,. \quad (219)$$

The contact terms are cancelled again. In these formulas $q_-^\alpha = (E_{\boldsymbol{q}}, -\boldsymbol{q})^\alpha$ and $k^\alpha = (\omega_{\boldsymbol{p}-\boldsymbol{q}}, \boldsymbol{p}-\boldsymbol{q})^\alpha$.

The nucleon mass shift due to the ρ-meson exchange can be found by substituting Eqs. (214), (216) and (217) to (213)

$$\delta m_N = I_1' + I_2'\,, \quad (220)$$

where

$$I_1' = \frac{1}{m(2\pi)^3}\int \frac{d\boldsymbol{q}}{E_{\boldsymbol{q}}}\frac{1}{(p-q)^2 - m_b^2}\Big\{g^2(2m^2 - pq)$$

$$+ 3gf(m^2 - pq) + \frac{f^2}{4m^2}(m^2 - pq)(5m^2 - pq)\Big\}, \quad (221)$$

$$I_2' = \frac{1}{2m(2\pi)^3} \int \frac{d\boldsymbol{k}}{\omega_{\boldsymbol{k}}} \frac{1}{(p+k)^2 - m^2} \Big\{ 2g^2(m^2 - pk) + 3gfm_b^2 + f^2(m_b^2 + \frac{(pk)^2}{2m^2}) \Big\}. \quad (222)$$

And the electron mass shift due to the photon exchange

$$\delta m_{e^-} = \frac{e^2}{8(2\pi)^3} \int \frac{d\boldsymbol{q}}{E_{\boldsymbol{q}} \omega_{\boldsymbol{p}-\boldsymbol{q}}} \sum_{\mu} \bar{u}(p\mu)\gamma_\alpha \times \Big\{ \frac{\not{q} + m}{E_{\boldsymbol{p}} - \omega_{\boldsymbol{p}-\boldsymbol{q}} - E_{\boldsymbol{q}}} + \frac{\not{q}_- - m}{E_{\boldsymbol{p}} + \omega_{\boldsymbol{p}-\boldsymbol{q}} + E_{\boldsymbol{q}}} \Big\} \gamma^\alpha u(p\mu) \quad (223)$$

is reduced to

$$\delta m_{e^-} = I_1'' + I_2'', \quad (224)$$

where

$$I_1'' = \frac{e^2}{m(2\pi)^3} \int \frac{d\boldsymbol{q}}{E_{\boldsymbol{q}}} \frac{2m^2 - pq}{(p-q)^2 - \lambda^2}, \quad (225)$$

$$I_2'' = \frac{e^2}{m(2\pi)^3} \int \frac{d\boldsymbol{k}}{\omega_{\boldsymbol{k}}} \frac{m^2 - pk}{(p+k)^2 - m^2}. \quad (226)$$

It is easy to see that the integrals (221) and (222) become equal, respectively, to (225) and (226) if we put $g = e$, $f = 0$ and $m_b = \lambda$.

Separate contributions in the curly brackets of (223) can be represented via the graphs (a) and (b) in Fig.18. The Fig.18 exemplifies a typical conversion of the non-relativistic propagators into Feynman-type ones. One can show that Eqs. (224)–(226) reproduce the one-loop integral for the corresponding self-energy contribution within the Dyson-Feynman formalism. More exactly, they coincide if one replace in the 4-dimensional integral the bare fermion mass and charge by their physical counterparts.

Conclusion

In many textbooks on nuclear physics we often encounter the Hamiltonian

$$H = K + V, \quad (227)$$

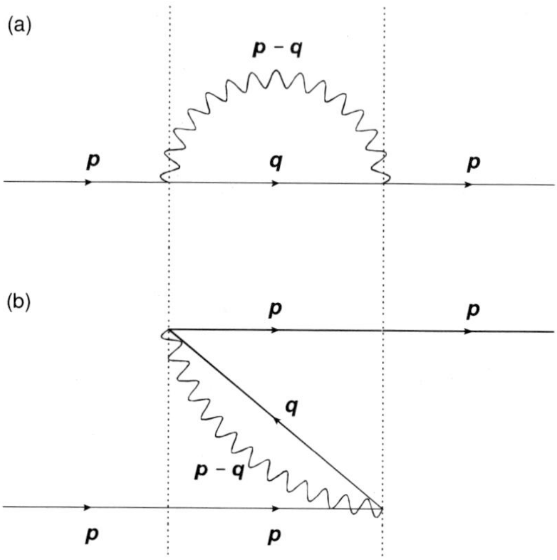

Figure 18. Two contributions to the mass shift within the Old-Fashioned Perturbation Theory.

where K the one-body operator of kinetic energy while the interactions between nucleons are written as

$$V = \sum_{i<j}^{N} V(i,j) + \sum_{i<j<k}^{N} V(i,j,k) + \cdots , \tag{228}$$

with two-body $V(i,j)$, three-body $V(i,j,k)$ forces, etc. $(i,j,k = 1,2,\cdots,N)$.

We have shown that the UCT method has allowed us to construct such forces on one and the same physical footing. And we have seen how along this guideline one can avoid some inconsistencies between the derivations of the expressions for $2N$- and $3N$- forces in available approaches to this problem.

Starting from the total Hamiltonian for interacting meson and nucleon fields, we come to the Hamiltonian and boost generator in the CPR, whose interaction parts consist of new relativistic interactions responsible for physical (not virtual) processes, particularly, in the system of bosons ($\pi-$, $\eta-$, $\rho-$, $\omega-$, $\delta-$ and $\sigma-$mesons) and fermions (nucleons and antinucleons).The corresponding quasipotentials (these essentially nonlocal objects) for the binary processes $NN \rightarrow NN$, $\bar{N}N \rightarrow \bar{N}N$, etc. are Hermitian and energy independent. It makes them attractive for various applications in nuclear physics. They embody the off-shell and recoil effects (the latter in all orders of the $1/c^2$ - expansion) without addressing any off-shell extrapolations of the $S-$matrix for the NN scattering.

In this work, we have shown a number of successful applications of the Kharkov potential built up in the second order in the coupling constants within the field theoretic approach. From the theoretical point of view, its further prospects are connected with constructing a new family of interactions between clothed particles (mesons and nucleons) in higher orders in the coupling constants (including three-body forces).

The triton binding energy has been calculated by solving the Faddeev $3N$ equations with the Kharkov potential. It results in the binding energy -7.799 MeV that is smaller than the experimental value (-8.48 MeV). Apparently, such a difference can be reduced after including $3N$-forces (perhaps, those derived here).

We have constructed a new family of relativistic energy independent interactions between the clothed electrons, positrons and photons. We have shown that the UCT method can be successfully applied to treatment of the bound states in QED. We have calculated the energy shift for the para-positronium ground state, which is surprisingly well coincides with those obtained earlier. Also, we have estimated the para-positronium decay rate into two photons. The obtained result is in a fair agreement with the experimental data.

By using a comparatively simple analytical means, we could show that the three-dimensional integrals, which determine the fermion mass renormalizations in the second order in the coupling constants, can be written in terms of the Lorentz invariants composed of the particle three-momenta. In other words, the corresponding renormalization integrals are independent of the particle momentum. The experience acquired has allowed us, on the one hand, to reproduce the manifestly covariant result obtained by Feynman techniques and, on the other hand, to derive a new representation for the Feynman integral that corresponds to the fermion self-energy diagram.

As a whole, persistent clouds of virtual particles are no longer explicitly contained in CPR, and their influence is included in properties of clothed particles (these quasiparticles of the UCT method). In addition, we would like to stress that problem of the mass and vertex renormalizations is intimately interwoven with constructing the interactions between the clothed particles. Renormalized quantities are calculated step by step in the course of the clothing procedure.

Acknowledgments

We are very grateful to J. Golak, H. Kamada, R. Skibiński, M. Stepanova and H. Witała for the fruitful collaboration that gives us an encouraging impetus for our research work. As before, we are also thankful to F. Gross and A. Stadler for sending the results of their analysis of modern np scattering data.

REFERENCES

[1] Al-Ramadhan, A. and D. Gidley (1994). New precision measurement of the decay rate of singlet positronium. *Phys. Rev. Lett. 72*, 1632.

[2] Arslanaliev, A., P. Frolov, J. Golak, H. Kamada, A. Shebeko, R. Skibiski, M. Stepanova, and H. Witaa (2021). Possible extension of the three-body force by the unitary clothing transformations method in the Faddeev equations. *Few Body Syst. 62*, 71.

[3] Arslanaliev, A., H. Kamada, A. Shebeko, M. Stepanova, H. Witala, and S. Yakovlev (2018). Applications of the Kharkov potential in the theory of nuclear forces and nuclear reactions. *Problems of Atomic Science and Technology 115*(3), 3–7.

[4] Arslanaliev, A., Y. Kostylenko, and A. Shebeko (2020). QED in the clothed-particle representation. In N. Orr et al. (Eds.), *Recent Progress in Few-Body Physics*, pp. 41. Springer.

[5] Belinfante, F. (1940). On the covariant derivative of tensor-undors. *Physica 7*, 305.

[6] Bjorken, J. and S. Drell (1964). *Relativistic Quantum Mechanics*. McGraw-Hill.

[7] Coon, S. A. and W. Glöckle (1981). Two-pion-exchange three-nucleon potential: Partial wave analysis in momentum space. *Phys. Rev. C 23*, 1790–1802.

[8] Dubovyk, I. and A. Shebeko (2010). The method of unitary clothing transformations in the theory of nucleon-nucleon scattering. *Few-Body Syst. 48*, 109.

[9] Frolov, P. and A. Shebeko (2015). *Relativistic interactions in meson-nucleon systems in the CPR*. Lambert Academic Publishing.

[10] Greenberg, O. and S. Schweber (1958). Clothed particle operators in simple models of quantum field theory. *Nuovo Cim. 8*, 378.

[11] Gross, F. and A. Stadler (2008). Covariant spectator theory of np scattering: Phase shifts obtained from precision fits to data below 350 mev. *Phys. Rev. C 78*, 014005.

[12] Kamada, H., A. Shebeko, and A. Arslanaliev (2017). Triton binding energy of Kharkov potential. *Few Body Syst. 58*, 70.

[13] Kamada, H., O. Shebeko, A. Arslanaliev, H. Witała, J. Golak, R. Skibiński, M. Stepanova, and S. Yakovlev (2020). Relativistic Faddeev Calculation for Nucleon-Deuteron Scattering with the Kharkov Potential. *Springer Proc. Phys. 238*, 449–453.

[14] Kleinert, H. (2016). *Particles and Quantum Fields*. Freie Universität Berlin (2016).

[15] Korchin, A. and A. Shebeko (1993). Ocubo effective operators method and relativistic model of nuclear structure. *Phys. At. Nucl. 56*, 1663.

[16] Korda, V., L. Canton, and A. Shebeko (2007). Relativistic interactions for the meson-two-nucleon system in the clothed-particle unitary representation. *Ann. Phys. 322*, 736.

[17] Korda, V. and A. Shebeko (2004). Clothed particles representation in quantum field theory: mass renormalization. *Phys. Rev. D 70*, 085011.

[18] Kostylenko, Y., A. Arslanaliev, and A. Shebeko (2020). QED in the clothed-particle representation: a fresh look at positronium properties treatment. In *SciPost Phys. Proc.* **3**, pp. 45.

[19] M. Fuda, Y. Z. (1995). Light front dynamics of one boson exchange models of the two-nucleon system. *Phys. Rev. C 51*, 23.

[20] Machleidt, R. (1989). The meson theory of nuclear forces and nuclear structure. *Adv. Nucl. Phys. 19*, 189.

[21] Machleidt, R., K. Holinde, and C. Elster (1987). The Bonn meson exchange model for the nucleon nucleon interaction. *Phys. Rep. 149*, 1.

[22] Mel'nik, Y. and A. Shebeko (1992). Calculation of proton polarization in deuteron disintegration with longitudinally polarized electrons. *Few Body Syst. 13*, 59.

[23] Mel'nik, Y. P. and A. V. Shebeko (1993). Electrodisintegration of polarized deuterons. *Phys. Rev. C 48*, 1259–1268.

[24] N. Bogoliubov, V. Tolmachev, D. S. (1958). A new method in the theory of superconductivity. *Fortschritte der Physik 6*, 605.

[25] Pirenne, J. (1946). The proper field and the interaction of Dirac particles. *Arch. Sci. Phys. Nat. 28*, 233.

[26] Sagara, K., H. Oguri, S. Shimizu, K. Maeda, H. Nakamura, T. Nakashima, and S. Morinobuet (1994). Energy dependence of analyzing power Ay and cross section for $p + d$ scattering below. *Phys. Rev. C 50*, 576.

[27] Sakai, H., K. Sekiguchi, H. Witała, W. Glöckle, M. Hatano, H. Kamada, H. Kato, Y. Maeda, A. Nogga, T. Ohnishi, H. Okamura, N. Sakamoto, S. Sakoda, Y. Satou, K. Suda, A. Tamii, T. Uesaka, T. Wakasa, and K. Yako (2000). Precise Measurement of dp Elastic Scattering at 270 MeV and Three-Nucleon Force Effects. *Phys. Rev. Lett. 84*, 5288–5291.

[28] Sakamoto, N., H. Okamura, T. Uesaka, S. Ishida, H. Otsu, T. Wakasa, Y. Satou, T. Niizeki, K. Katoh, T. Yamashita, K. Hatanaka, Y. Koike, and H. Sakai (1996). Measurement of the vector and tensor analyzing powers for the d-p elastic scattering at Ed = 270 MeV. *Physics Letters B 367*(1), 60–64.

[29] Schweber, S. (1961). *An Introduction to Relativistic Quantum Field Theory*. New York: Row, Peterson & Co.

[30] Shebeko, A. (2002). In *PoS Baldin ISHEPP XVI*, Dubna, Russia.

[31] Shebeko, A. (2004). The S-matrix in the method of unitary clothing transformations. *Nucl. Phys. A 737*, 252.

[32] Shebeko, A. (2012). The method of unitary clothing transformations in quantum field theory: Applications in the theory of nuclear forces and reactions. In S. Ketov (Ed.), *Advances in quantum field theory*, Chapter 1, pp. 1. INTECH.

[33] Shebeko, A. (2016, 01). Clothed particles in quantum electrodynamics and quantum chromodynamics. *EPJ Web of Conferences 113*, 03014.

[34] Shebeko, A. and I. Dubovyk (2013). Low-energy parameters of the nucleonnucleon scattering and deuteron properties, electromagnetic interactions with bound systems. *Few-Body Syst. 54*, 1513.

[35] Shebeko, A. and P. Frolov (2012). A possible way for constructing generators of the Poincaré group in quantum field theory. *Few-Body Syst. 52*, 125.

[36] Shebeko, A. and M. Shirokov (2000). Clothing procedure in relativistic quantum field theory and its applications to description of electromagnetic interactions with nuclei (bound systems). *Progr. Part. Nucl. Phys. 44*, 75.

[37] Shebeko, A. and M. Shirokov (2001). Unitary transformations in quantum field theory and bound states. *Phys. Part. Nucl. 32*, 15.

[38] Shirokov, M. and M. Visinescu (1974). Perturbation approach to the field theory, "dressing" and divergences. *Rev. Roum. Phys. 19*, 461.

[39] Smith, C. (2002). *Bound State Description in Quantum Electrodynamics and Chromodynamics*. Ph. D. thesis, Louvain-la-Neuve.

[40] Tamura, K., T. Niwa, T. Sato, and H. Ohtsubo (1988). Exchange current due to rho-meson exchange. *Prog. Theor. Phys. 80*, 138.

[41] Weinberg, S. (1995). *The Quantum Theory of Fields*, Volume 1. Cambridge University Press.

[42] Wheeler, J. (1946). Polyelectrons. *Ann. N. Y. Acad. Sci. 48*, 219.

In: Quantum Field Theory and Applications
Editor: N. Palerma
ISBN: 978-1-68507-930-7

Chapter 2

THE QUANTUM FIELD THEORY IN DISCRETE AND CONTINUOUS PICTURE, QUANTUM FOAM AND ULTRAVIOLET DIVERGENCE PROBLEM

Alexander E. Shalyt-Margolin*
Institute for Nuclear Problems, Belarusian State University, Minsk, Belarus

Abstract

Based on the results from black hole thermodynamics at all energy scales, this work demonstrates that, both for the discrete QFT previously introduced by the author and for QFT in continuous space-time, there is a natural ultraviolet applicable boundary (cut-off) distant from the Planck scales. It is important that this boundary exists irrespective of the fact in which pattern, perturbative or non-perturbative mode, QFT is studied. Different inferences from the obtained results are discussed, some statements are revised.

Keywords: quantum field theory, quantum foam, ultraviolet divergence problem

PACS: 11.10.-z,11.15.Ha,12.38.Bx

---

*Corresponding Author's Email: a.shalyt@mail.ru; alexm@hep.by.

1. Introduction

This paper redefines and supplements the results obtained by the author in [1]–[10].

Section 2 of the paper presents a brief review of the previously obtained results from the works [1]–[9] which contain detail studies of a discrete QFT at low energies based on the existence of the maximal momentum $E \ll E_p$ at the Planck's level $p_{max} \approx P_{pl}$ and of the corresponding length $\ell \doteq \hbar/p_{max}$ that is named *primary length*. In [1]–[9] the discrete theories based on the existence of p_{max} and $\ell \doteq \hbar/p_{max}$ are referred to as *measurable* theories, the mathematical apparatus is formed, their common properties are found. Besides, it is shown that at low energies $E \ll E_p$ they are very close to the initial theories in continuous space-time.

Section 3 the correlation between the *measurability* and the lattice approach in QFT is revealed. In fact, it is shown that measurable theories can be regarded as lattice theories. It is shown that measurable theories can be viewed as lattice theories with very small lattice spacing.

Section 4 is central in this presentation. In its first part of Section 4 the principal result associated with the scalar model φ^4 from [8] is elucidated. In the second part of this section it is shown that the methods and results presented in the first part may be generalized to the quantum theory of gauge fields interacting with fermion fields of the matter.

Section 5 presents the formulation of the hypothesis (*Main Hypothesis*) concerning the applicability boundary of QFT both in continuous space-time and in the case of discrete consideration. It is noted that, provided the *Main Hypothesis* is true, it should be valid without regard to perturbative or nonperturbative mode of QFT studies.

In Section 6 on the basis of the results obtained in [10], the above-mentioned hypothesis is proved within a natural assumption that at Planck's scales the approach $l \approx l_p$ *quantum foam* introduced in the mid-fifties of the past century by J.A.Wheeler involves the Schwarzschild micro-black holes **mbh** with the radius $r = r_{mbh}$ on the order of the Planck length $r_{mbh} \propto l_p$ and with the mass $m = m_{mbh}$ on the order of the Planck mass $m_{mbh} \propto m_p \approx 10^{-5}g$. As compared to [10], Section 6 presents important revisions and supplements. Finally, Section 7 is devoted to the inferences from the obtained results and to refinement of some statements.

2. Measurability Concept in Quantum Theory and Gravity

In this Section we briefly consider some of the results from [1]–[7] which are essential for subsequent studies. Without detriment to further consideration, in the initial definitions we lift some unnecessary restrictions and make important specifications.

Presently, many researchers are of the opinion that at very high energies (Plank's or trans-Planck's) the ultraviolet cutoff exists that is determined by some maximal momentum.

Therefore, it is further assumed that there is a maximal bound for the measurement momenta $p = p_{max}$ represented as follows:

$$p_{max} \doteq p_\ell = \hbar/\ell, \tag{1}$$

where ℓ is some small length and $\tau = \ell/c$ is the corresponding time. Let us call ℓ the *primary* length and τ the *primary* time.

Without loss of generality, we can consider ℓ and τ at Plank's level, i.e., $\ell \propto l_p, \tau = \kappa t_p$, where the numerical constant κ is on the order of 1. Consequently, we have $E_\ell \propto E_p$ with the corresponding proportionality factor, where $E_\ell \doteq p_\ell c$.

Explanation. In the theory under study it is not assumed from the start that there exists some minimal length l_{min} and that ℓ is such. In fact, the minimal length is defined with the use of Heisenberg's Uncertainty Principle (HUP) $\Delta x \cdot \Delta p \geq \frac{1}{2}\hbar$ or of its generalization to high (Planck) energies – Generalized Uncertainty Principle (GUP) [11]–[14], for example, of the form [11]

$$\Delta x \geq \frac{\hbar}{\Delta p} + \alpha' l_p^2 \frac{\Delta p}{\hbar}, \tag{2}$$

where α' is a constant on the order of 1. Evidently this formula (2) initially leads to the minimal length $\widetilde{\ell}$ on the order of the Planck length $\widetilde{\ell} \doteq 2\sqrt{\alpha'} l_p$. Besides, other forms of GUP [14] also lead to the minimal length.

Thus, we should note that in all the works l_{min} is actually (but not explicitly) introduced on the basis of some measuring procedure (different forms of the Generalized Uncertainty Principle (GUP)). In any form GUP in turn is a high-energy generalization of HUP. But in the original proof of HUP a planar geometry of the initial space-time was actively used [15]. Extension of this principle to other pairs of conjugate variables is also valid only for quantum mechanics in the planar geometry space [16]. As HUP is a local principle, at low energies in the curved space-time, by virtue of Einstein's Equivalence Principle, we can consider that in a fairly small neighborhood of any point the geometry is planar an hence HUP is valid too. But all the results obtained point to the fact that l_{min} should be at a level of l_p,i.e., $l_{min} \propto l_p$, or even should be smaller. As noted in the Section 2 of [7], at the Planck scales Einstein's Equivalence Principle is obviously inapplicable, and there is no way to use the measuring procedure ignoring the space geometry at these scales. Meantime, none of the GUP forms [14] makes an effort to include it and is hardly completely correct. Moreover, there are some serious arguments against GUP as demonstrated in Section IX of the review paper [14]. The foregoing considerations support argumentation against the introduction of l_{min} from the start.

Because of this, in the present work the validity of this principle is not implied from the start too. GUP is given merely as an example. As p_{max} (1) is taken at Planck's level, it is clear that HUP is inapplicable. Taking this into consideration, the existence of a certain minimal length $\widetilde{\ell}$ is not mandatory. So, we start from the *primary* length ℓ and the *primary* time τ. The whole formalism, developed in [1]–[6] on condition that ℓ is the minimal length, is valid for the case when ℓ is the *primary* length but now we can lift the formal requirement for involvement of l_{min} in the theory from the start.

There is one more barrier for the use of l_{min} in the theory as indicated in [12] and other works (for example, [14]). In the above-mentioned papers, it has been noted that there is a nonzero minimal uncertainty in position, i.e., l_{min} implies that there is no physical state which is a position eigenstate since an eigenstate would, of course, have zero uncertainty in position. So, in this case in a quantum theory we have the momentum representation rather than the position representation, and the quantum theory becomes very depleted.

The question arises whether the introduction of p_{max} is naturally associated with the involvement of a minimal length. But this is the case only when at the energies E_{max} corresponding to p_{max} we have the substantiated measuring procedure. Unfortunately, this is not the case.

Note that in the canonical QFT in continuous space-time (i.e., without l_{min})[17]–[20] measurements of the contributions in the loop amplitudes involve the standard cut-off procedure for some large (maximal) momentum $p_{cut} \doteq p_{max}$. Then it is demonstrated that the theory at low energies $p \ll p_{cut}$ is in fact independent of the selection of $p_{cut} \doteq p_{max}$. Of course, the theory still remains to be continuous [17]–[20]. In this case we make another step forward, relating the corresponding length $\ell = \hbar/p_{max}$ to p_{max} and constructing on its basis a low-energy theory very close to the initial continuous theory. Now we have the naturally derived parameter ℓ for the construction of a high-energy deformation of this theory at the energies $E \approx E_{max}$ within the scope of determining the physical theory deformation [21]. So, we start from the *primary* length ℓ and the *primary* time τ. The whole formalism, developed in [1]–[6] on condition that ℓ is the minimal length, is valid for the case when ℓ is the *primary* length but now we can lift the formal requirement for involvement of l_{min} in the theory from the start.

In what follows we mainly make references to [6] and [7]. In particular, the basic definitions rimary Measurability, Generalized Measurability, Primarily Measurable Quantities (PMQ), Primarily Measurable Momenta (PMM), Generalized Measurable Quantities (GMQ) and the like are given in Section II of [6].

Besides, in Section III from [6] it has been demonstrated how, at low energies $E \ll E_p$, the arbitrary metric $g_{\mu\nu}(x)$ may be derived in terms of *measurable* quantities.

It should be noted that in virtue of assumption in [7] *observables* in *measurable* theory are *Primarily Measurable Quantities*.

3. QFT in Measurable Form as Lattice Field Theory

From Section 4 in [7] it follows directly that the *measurable* approach generates a Lattice Quantum Field Theory (LQFT). Hereinafter we use the symbols, terms, and results from the (LQFT) [22],[23].

Then it is assumed that the theory under study is considered in a sufficiently large hypercubic box with the edge length L and space-time size L^4, where $L = N_L\ell$, $N_L \gg 1$. In general, not necessarily N_L is an integer number.

For convenience, let us introduce the following:

$$\Omega \doteq L^4. \tag{3}$$

Assuming that t varies over the interval $0 \leq t \leq T, T \neq L$, (3) will take the form (formula (2.3) in [23])

$$\Omega \doteq VT = L^3T. \tag{4}$$

In what follows, when not stated otherwise, we assume $T = L$ and hence formula (4) takes the form (3).

Without loss of generality, we can assume that all integer N_{x_μ} are equal to each other and are equal to some integer $N \gg 1$ maximally high in the absolute value. Then, according to the present consideration, in the *measurable* form there arises a lattice model of the position representation with $a = \ell/N$, where a is the lattice distance or same lattice spacing (section 2.5 from [22]).

In line with the general approach, in LQFT we have [22]

$$L = aM = \frac{\ell}{N}M, i.e., \frac{L}{M} = \frac{\ell}{N}, \tag{5}$$

where $M \gg 1$ is an integer number. It is obvious that $M/N = N_L$, i.e., $M \gg N_L$.

As L is great, also without loss of generality, it is assumed that the periodic boundary conditions (formula (2.58) in [22])) are valid

$$\phi(x+L) = \phi(x). \tag{6}$$

Then all formulae of LQFT in the position representation (Sections 2.5,2.6 in [22]) are valid for the *measurable* form of a continuous theory. And formula (2.54) from [22]

$$\sum_x f(x) \rightarrow \int_0^L d^4x f(x), M \rightarrow \infty, a = \frac{L}{M}, L \quad fixed \tag{7}$$

may be rewritten for such consideration with substitution of $f(x) \rightarrow \mathcal{L}$ under the integration sign for $f(x) \rightarrow \mathcal{L}_{meas,\{N\}}$ within the summation, and $a \rightarrow \ell/N$, where $\mathcal{L}$ and $\mathcal{L}_{meas,\{N\}}$ are correspondent formulae in continuous and measurable cases.

Since ℓ is also a fixed quantity, it is clear that the conditions $M \rightarrow \infty$ and $N \rightarrow \infty$ in the case under study are equivalent, representing the thermodynamic limit that gives a continuous pattern. Note that in this case we can use the results from Sections 2.5 and 2.6 of [22],assigning a_t as the temporal lattice distance $a_t \doteq \tau/N_t$, where τ/N_t is taken from formula (4) in [6].

Thus, in the coordinate representation the studied lattice of *measurable* quantities may be regarded as a canonical space-time lattice of LQFT, with the spacing $a = \ell/N$ and temporal distance $a_t = \tau/N_t$.

In this case all the basic operators in Sections 2.5 and 2.6 of [22] have their analogs in the present work. Specifically, finite-differences operators

$\partial_\mu \varphi_x, \partial'_\mu \varphi_x$ from formulae (2.55),(2.56) in [22] and formulae of Section 2.3 in [24] in the present paper correspond to the operators $\frac{\Delta}{\Delta_N}$ from formula (9) in [6] for positive and negative values of **N**. The transfer-operator $\hat{T}$ may be constructed for the lattice of interest, with the spacing $a = \ell/N$ and temporal distance $a_t = \tau/N_t$, in accordance with formulae (2.71),(2.74) of [22], so all the formulae from Section 2.6 in [22] are valid for this case. We assume that $a_t = a$.

For the lattice values of momenta, in the momentum representation, according to formula (2.81) in [22], we have

$$p_\mu(latt) = n_\mu \frac{2\pi}{L}, \tag{8}$$

where n_μ are integers.

Consequently, the lattice edge in the momentum representation $\Delta p_\mu(latt)$ adopts the value

$$\Delta p_\mu(latt) = \frac{2\pi}{L} \propto \frac{1}{N_L}, \tag{9}$$

where it is assumed that $\hbar = 1$.

At the same time, the integer numbers n_μ are varying in magnitude over the interval $[0, N_L N]$,, where $N_L N = L/a$ (formula (2.82) in [22]). As a result, in the case of interest a maximum value of the momentum along any axis will be given by

$$p_{latt,max} = \frac{\pi}{a} = \frac{\pi}{\ell/N} = \frac{\pi N}{\ell} \doteq \Lambda. \tag{10}$$

Formula (10) gives an explicit expression for a maximal lattice momentum $p_{latt,max} = \Lambda$. To be more exact, the momenta are restricted to the so-called first Brillouin zone (**BZ**) $\mathcal{B}$ (formula (1.218) from [23])

$$\mathcal{B} \doteq \{p| \frac{-\pi}{a} < p_\mu \leq \frac{\pi}{a}\}. \tag{11}$$

It is clear that $p_{latt,max} = \Lambda \gg p_\ell$. As follows from formula (10), $\Lambda \propto N p_\ell, N \gg 1$,i.e., the boundary of **BZ** Λ passes far beyond the region of the physical energy values.

But due to the condition $E \ll E_p$, we consider only a low-energy part of the lattice, the momenta of which are given as $p \approx \frac{\hbar}{N^* \ell}$ with $|N^*| \gg 1$. Because of this, in the case under study only particular momenta may be maximal (so-called "maximally reachable" momentum) $p_{max,reach}$ and $p_{max,reach} \ll p_{latt,max}$.

In this way **BZ** in formula (11) is narrowed significantly

$$-p_{max,reach} \leq p_{\mu} \leq p_{max,reach}, \tag{12}$$

where $p_{max,reach} \ll p_{\ell}$.

As $a = \ell/N$,where $N \gg 1$, when the mass m is fixed, am is close to zero and hence the correlation length ξ (formula (1.224) in [23])

$$\xi \equiv \frac{1}{am} = \frac{N}{\ell m} \tag{13}$$

is finite but very great. Passage to a continuum limit $\xi \to \infty$ means going to $N \to \infty$. In this case, within the constant factor m^{-1}, we have

$$\xi = \frac{N}{\ell} \propto N p_{\ell} \approx N p_{pl} \propto p_{latt,max} = \Lambda. \tag{14}$$

From formulae (10),(12) it follows directly that

$$p_{max,reach} = \frac{p_l}{\tilde{N}} = \frac{\Lambda}{N\tilde{N}}, N \gg 1, \tilde{N} \gg 1. \tag{15}$$

Then, proceeding from the formulae above, in the case of interest (**BZ**) $\mathcal{B}$ (11) is narrowed to $\mathcal{B}_{\mathcal{N}}$

$$\mathcal{B}_{\mathcal{N}} \doteq \{p|\frac{-\pi}{N\tilde{N}a} < p_{\mu} \leq \frac{\pi}{N\tilde{N}a}\}, N \gg 1, \tilde{N} \gg 1. \tag{16}$$

Lattice summation in the general case is given by formula (2.7) from [23]

$$\int_{p\in\mathcal{B}} \doteq \int_{\mathcal{B}} \equiv \frac{1}{a^4\Omega}\sum_{p\in\mathcal{B}}. \tag{17}$$

In the case under study the lattice summation takes the form

$$\int_{p\in\mathcal{B}_{\mathcal{N}}} \doteq \int_{\mathcal{B}_{\mathcal{N}}} \equiv \frac{1}{a^4\Omega}\sum_{p\in\mathcal{B}_{\mathcal{N}}}. \tag{18}$$

Respectively, on passage to the thermodynamic limit $L \to \infty, T \to \infty$, in the general case we arrive at formula (2.8) in [24]

$$\int_{p \in \mathcal{B}} = \frac{1}{(2\pi)^4} \int_{\frac{-\pi}{a}}^{\frac{\pi}{a}} d^4 p. \tag{19}$$

In the case of interest (19) is transformed to

$$\int_{p \in \mathcal{B}_\mathcal{N}} = \frac{1}{(2\pi)^4} \int_{\frac{-\pi}{N\tilde{N}a}}^{\frac{\pi}{N\tilde{N}a}} d^4 p. \tag{20}$$

Remark 3.1. As a rule, in the literature devoted to LQFT it is assumed that the lattice edge a is equal to 1. Then the formula for the first Brillouin zone $\mathcal{B}$ (11) is of the form

$$\mathcal{B} \doteq \{p| -\pi < p_\mu \leq \pi\}. \tag{21}$$

Whereas for the "short-cut" Brillouin zone $\mathcal{B}_\mathcal{N}$ (16) we have

$$\mathcal{B}_\mathcal{N} \doteq \{p|\frac{-\pi}{N\tilde{N}} < p_\mu \leq \frac{\pi}{N\tilde{N}}\}, N \gg 1, \tilde{N} \gg 1, \tag{22}$$

with the corresponding changes in all other formulae.

4. Perturbation Theory in Continuous and Measurable Cases

4.1. Simple Scalar Model φ^4

The canonical Lagrangian for model φ^4 in continuous space-time has the form [17]

$$\mathcal{L} = \frac{1}{2} (\partial_\mu \varphi)^2 - \frac{m_0^2}{2} \varphi^2 - \frac{\mathrm{g}_0}{4!} \varphi^4 \tag{23}$$

where $\mathcal{L}_0 \doteq \frac{1}{2}((\partial_\mu \varphi)^2 - m_0^2 \varphi^2)$ is free fields Lagrangian and $\mathcal{L}_I \doteq -\frac{\mathrm{g}}{4!}\varphi^4$ is interaction Lagrangian and g is a dimensionless constant (in four dimensions).

Using in measurable form operator $\frac{\Delta}{\Delta_{N_{x\mu}}}$ from formula (9) in [6], we can easily obtain, instead of $\mathcal{L}$ its *measurable* form

$$\mathcal{L}_{meas,\{N\}} = \frac{1}{2}(\frac{\Delta}{\Delta_{N_{x\mu}}}\varphi_{meas})^2 - \frac{1}{2}m_0^2\varphi_{meas}^2 - \frac{g_0}{4!}\varphi_{meas}^4 \tag{24}$$

and instead $\mathcal{L}_0$ with the corresponding *Klein–Gordon equation* or *KGE*

$$(\Box + m_0^2)\,\phi = 0 \tag{25}$$

their *measurable* forms

$$\mathcal{L}_{meas,\{N\},0} = \frac{1}{2}(\frac{\Delta}{\Delta_{N_{x\mu}}}\phi_{meas})^2 - \frac{m_0^2}{2}\,\phi_{meas}^2 \tag{26}$$

and

$$(\Box_{\mathbf{N_{x\mu}}} + m_0^2)\,\phi_{meas} = 0\,. \tag{27}$$

Within the scope of a perturbation theory, let us consider examples of Feynman diagrams, which give UVD for the φ^4-model in canonical QFT in continuous space-time [17] – [20], to find what are the correspondences with a *measurable* picture.

Now we consider one-loop corrections for the two- and four-vertex functions:

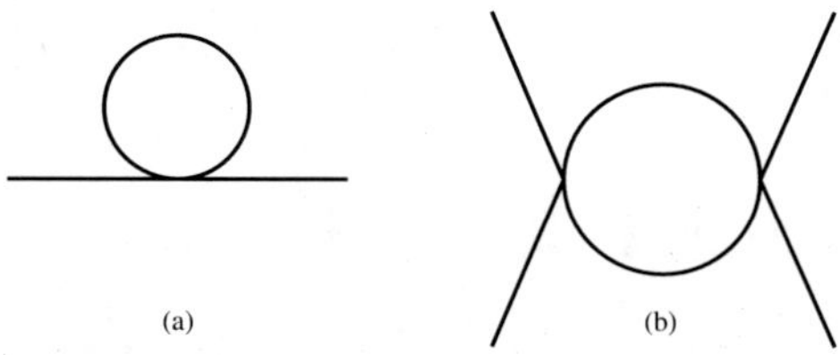

Figure 1. Diagrams (a) and (b).

Then the quantity $G(0)$, quadratically divergent over the momentum k (associated with the diagram (a) in Fig.1, formula (9.1) in [17])

$$G(0) = g_0 \int \frac{d^4k}{(2\pi)^4}\,\tilde{G}(k) = g_0 \int \frac{d^4k}{(2\pi)^4}\frac{1}{k^2 - m_0^2} \tag{28}$$

corresponds in a *measurable* picture to the integral, finite over k with $|N^*| = \infty$

$$G(0, N_*) \doteq g_0 \int\limits_{-p_{N_*}}^{p_{N_*}} \frac{d^4k}{(2\pi)^4} \frac{1}{k^2 - m_0^2}. \tag{29}$$

Similarly, another divergent diagram–graph of the order $O(g^2)$, whose contribution is represented by the logarithmically divergent integral (formula (9.2) in [17])

$$g_0^2 \int \frac{d^4k}{(2\pi)^8} \frac{1}{(k^2 - m_0^2)((p_1 + p_2 - k)^2 - m_0^2)} \tag{30}$$

in a *measurable* consideration will be associated with the finite quantity

$$g_0^2 \int\limits_{-p_{N'_*}}^{p_{N'_*}} \frac{d^4k}{(2\pi)^8} \frac{1}{(k^2 - m_0^2)((p_1 + p_2 - k)^2 - m_0^2)}. \tag{31}$$

(Here in the *measurable* case the right-hand sides of formulae (29),(31) should have the corresponding sums instead of the integrals but,in virtue the final part of Section 3 the sums may be replaced by the corresponding integrals).

It should be noted that we can pass to Euclidean space-time by means of Wick rotation (Remark 2.4) for better convergence of the integrals. Then, with the help of an analytical extension, we can return to Minkowskian space-time. This is a standard method both for QFT and LQFT [22], [23]. The continuum action of the theory (23) in Euclidean space-time is of the form (formula (2.17) from [24])

$$S = \int d^4x (\frac{1}{2} (\partial_\mu \varphi)^2 + \frac{m_0^2}{2} \varphi^2 + \frac{g_0}{4!} \varphi^4), \tag{32}$$

and the corresponding lattice action has the following form:

$$S_{meas,\{N\}} = a^4 \sum_x (\frac{1}{2} (\frac{\Delta}{\Delta_N} \varphi_{meas})^2 + \frac{1}{2} m_0^2 \varphi_{meas}^2 + \frac{g_0}{4!} \varphi_{meas}^4). \tag{33}$$

For the lattice values of momenta, in the momentum representation take place the formulae (8),(9). However, here the difficulty arises – the corresponding lattice in the momentum representation on L^4 is uniform with the lattice spacing in formula (9).

In the considered case the lattice of *measurable* momenta is nonuniform with the lattice spacing

$$\triangle p_\mu(meas) = \frac{1}{(N^* - \kappa)(N^* - \kappa \mp 1)\ell}, \tag{34}$$

where κ is an integer number, $|\kappa| \ll |N^*| \gg 1$.

As shown in [7], in order to use the results from [22], it is required that the condition

$$\triangle p_\mu(latt) \approx \triangle p_\mu(meas) \tag{35}$$

be fulfilled.

As follows from formula (34) and [7] this is the case when

$$N_L \approx (N^*)^2. \tag{36}$$

This condition is quite natural considering that L may be chosen no matter how large but finite.

Now in the same way we consider the momentum representation and Fourier transformation of the above mentioned lattice (formula (1.171) in [23])

$$G(x-y;a) = \int\limits_{-\pi/a}^{\pi/a} \frac{d^4p}{(2\pi)^4} e^{ip(x-y)} \tilde{G}(p;a) = \int\limits_{p\in\mathcal{B}} \frac{d^4p}{(2\pi)^4} e^{ip(x-y)} \tilde{G}(p;a). \tag{37}$$

Then we can use the results of [23] to find, how well a continuous propagator of the momentum representation is approximated by the "lattice" propagator in this representation. As it has been noted, all calculations in [23] are first performed in Euclidean space-time and followed by the analytical extension to Minkowskian space.

In virtue, using formula (1.173) from [23], we have

$$\tilde{G}(p;a) = \{\sum_{\mu=1}^{4} a^{-2} 4 \sin^2 \frac{ap_\mu}{2} + m^2\}^{-1}. \tag{38}$$

But it has been shown that in the case under study the momenta p are taken only from the subset $\mathcal{B}_{\mathcal{N}}$. Consequently, $p_\mu \propto 1/N_\mu, |N_\mu| \gg 1$. As $a = \ell/N, N \gg 1$, the argument of the function $\sin^2$ is $\propto 1/(NN_\mu)$, i.e., it is very close to zero. Further we use a simple property – $\sin x \approx x$ for x close to 0. Immediately, within a high accuracy, by formula (38) we can obtain

$$\tilde{G}(p;a) = \{\sum_{\mu=1}^{4} a^{-2} 4 \frac{a^2 p_\mu^2}{4} + m^2\}^{-1} = \{\sum_{\mu=1}^{4} p_\mu^2 + m^2\}^{-1} = (p^2 + m^2)^{-1} \tag{39}$$

in a good agreement with the corresponding formula in a continuous picture, i.e., for $a \rightarrow 0$ ([23], formula (1.178)).

Thus, these computations once again demonstrate that in a *measurable* form at low energies $E \ll E_p$ the theory studied is to a high accuracy coincident with the corresponding theory in the continuous case.

Perturbation theory and Feynman rules for present lattice are analogous to a continuous theory but they have the interaction term

$$S^{(1)} = a^4 \sum_x \frac{g_0}{4!} \varphi_{meas}^4. \tag{40}$$

As distinct from a continuous consideration, by the lattice approach all Feynman graphs satisfy the following properties in momentum space ([23],p.64) in the general case:

each line is associated with the propagator $\tilde{\Delta}(q) \equiv (m_0^2 + \hat{q}^2)^{-1}$;

each vertex is an end point of four lines and is associated with the factor $-g_0$;

at in inner vertices momentum conservation holds modulo 2π;

loop momenta should to be integrated over the first Brillouin zone $\mathcal{B}$ with the integration measure $\int\limits_{p \in \mathcal{B}}$;

there is in overall factor $(2\kappa)^{-n/2}$ resulting from our normalization of the lattice scalar fields

formally UVD appear only in the continuum limit, i.e., when $a \rightarrow 0$.

Note that in the second point $-g_0$ should be replaced by $-g_{0,\mathcal{B}_{\mathcal{N}}}$, and it seems that the fourth item should be replaced by

loop momenta should be integrated over the short-cut Brillouin zone,$\mathcal{B}_{\mathcal{N}}$ with the integration measure $\int\limits_{p\in\mathcal{B}_{\mathcal{N}}}$.

As, for $N \gg 1$, the lattice edge $a = \ell/N$ is very small and hence the correlation length ξ (formula (13)) is very great but not infinite, the indicated lattice in the space-time and momentum representation is actually not distinct from a continuous consideration for the momenta satisfying **BZ** $\mathcal{B}$ (11).

Thus, as directly follows from formula (16), we should include the contributions made *only* by very small momenta p in $\mathcal{B}$,i.e., for $p \in \mathcal{B}_{\mathcal{N}}$. Taking this into account, further we use the known formulae of LQFT for small momenta (Section 2 in [23])).

We assume that the field $\varphi(x)$ in a *symmetric phase*

$$\langle\varphi(x)\rangle = 0, \tag{41}$$

i.e., Z_2-symmetry of $\varphi(x) \mapsto -\varphi(x)$ is the case, whereas Green's functions with an odd number of arguments vanish.

As it has been correctly noted in Section 2 of [18]:

> "...Renormalization has its own intrinsic physical basis and is not brought about solely by the necessity to expurgate infinities. Even in a totally finite theory we would still have to renormalize physical quantities".

This is associated with the fact that the theoretical initial (*bare*) quantities (mass m_0,charge q_0 and so on) can differ drastically from the real (physical) quantities (m_R,q_R and so on).But because in this case in the *measurable* picture at energies $E \ll E_p$ a low-energy part of the lattice is involved, very close to continuous space-time, there is a possibility to derive QFT without infinities, when renormalization of the theory is understood as a passage from some finite quantities to the other.

Next, we present briefly the results from [8].

In the general case a one-loop correction to the two-vertex function (diagram (a)) takes the form ([24],p.53):

$$\Gamma^{(2)}(p,-p) = -(\hat{p}^2 + m_0^2) - \frac{g_0}{2} J_1(m_0) \equiv$$
$$\equiv -(\hat{p}^2 + m_R^2), \tag{42}$$

where, as a rule, the term $\mathcal{O}(g_0^2)$ in the right-hand side is omitted and the designations from Section 2 in [23] are used: $\tilde{\Delta}(q) \equiv (m_0^2 + \hat{q}^2)^{-1}$, $J_n(m_0) \equiv \int\limits_{\mathcal{B}(q)} \tilde{\Delta}(q)^n$. Here where m_R is the renormalized mass in the general case and $\mathcal{B}(\tilde{q})$ is (**BZ**) for the variable $\tilde{q}$.

But, proceeding from the earlier results, in considered case it follows that $\Gamma^{(2)}(p,-p)$ should be replaced by

$$\Gamma^{(2)}(p,-p,\mathcal{B}_{\mathcal{N}}) = -(\hat{p}^2 + m_{0,\mathcal{B}_{\mathcal{N}}}^2) - \frac{g_{0,\mathcal{B}_{\mathcal{N}}}}{2} J_1(m_0,\mathcal{B}_{\mathcal{N}}) \equiv -(\hat{p}^2 + m_{R,\mathcal{B}_{\mathcal{N}}}^2), \tag{43}$$

where $p \in \mathcal{B}_{\mathcal{N}}$,

$$J_n(m_0,\mathcal{B}_{\mathcal{N}}) \equiv \int\limits_{\mathcal{B}_{\mathcal{N}}(q)} \tilde{\Delta}(q)^n, \tag{44}$$

and $m_{0,\mathcal{B}_{\mathcal{N}}}$, $g_{0,\mathcal{B}_{\mathcal{N}}}$ – corresponding *bare* mass and coupling constant within $\mathcal{B}_{\mathcal{N}}$. Here, $\mathcal{B}_{\mathcal{N}}(\tilde{q})$ is the narrowed (**BZ**) $\mathcal{B}_{\mathcal{N}}$ for the variable $\tilde{q}$, and in the right side (43) there is no term $\mathcal{O}(g_{0,\mathcal{B}_{\mathcal{N}}}^2)$ and $m_{R,\mathcal{B}_{\mathcal{N}}}$ are the experimental values of mass obtained for the energies on the order of $\mathcal{B}_{\mathcal{N}}$. Naturally, we can suppose that the renormalized (i.e., experimental) values of mass m_R and coupling constant g_R at energies $E \ll E_p$ *should not* depend on the whole domain of $\mathcal{B}$, the limiting values of which are much greater than E_p. Besides, in any region satisfying the condition $E \ll E_p$ they are independent of this domain and hence we have $m_{R,\mathcal{B}_{\mathcal{N}}} = m_R$, $g_{R,\mathcal{B}_{\mathcal{N}}} = g_R$.

In virtue of the condition $m_{R,\mathcal{B}_{\mathcal{N}}} = m_R$ and considering the terms $\mathcal{O}(g_0^2)$, $\mathcal{O}(g_{0,\mathcal{B}_{\mathcal{N}}}^2)$, we can rewrite formula (42) as (formula (2.93) in [23])

$$m_R^2 = m_0^2 + \frac{g_0}{2} J_1(m_0) + \mathcal{O}(g_0^2), \tag{45}$$

and formula (43) as

$$m_{R,\mathcal{B}_{\mathcal{N}}}^2 = m_R^2 = m_{0,\mathcal{B}_{\mathcal{N}}}^2 + \frac{g_{0,\mathcal{B}_{\mathcal{N}}}}{2} J_1(m_{0,\mathcal{B}_{\mathcal{N}}},\mathcal{B}_{\mathcal{N}}) +$$
$$+\mathcal{O}(g_{0,\mathcal{B}_{\mathcal{N}}}^2). \tag{46}$$

Similar calculations may be performed for the coupling constant too. Specifically, let $\Gamma_R^{(4)}(p_1, p_2, p_3, p_4)$ be the renormalized four-point function. Then, for the renormalized coupling constant g_R, we have ([23],formula (2.96))

$$g_R = -\Gamma_R^{(4)}(0,0,0,0) = g_0 - \frac{3}{2}g_0^2 J_2(m_0) + \mathcal{O}(g_0^3), \tag{47}$$

And, since $g_{R,\mathcal{B}_\mathcal{N}} = g_R$, we have

$$\begin{aligned} g_{R,\mathcal{B}_\mathcal{N}} = g_R = -\Gamma_{R,\mathcal{B}_\mathcal{N}}^{(4)}(0,0,0,0) = \\ = g_{0,\mathcal{B}_\mathcal{N}} - \frac{3}{2}g_{0,\mathcal{B}_\mathcal{N}}^2 J_2(m_{0,\mathcal{B}_\mathcal{N}}, \mathcal{B}_\mathcal{N}) + \mathcal{O}(g_{0,\mathcal{B}_\mathcal{N}}^3). \end{aligned} \tag{48}$$

As follows from the four last equations, since left sides of each pair of these equations are equal, whereas the integrals $J_1(m_0)$ and $J_1(m_0, \mathcal{B}_\mathcal{N})$ and hence $J_2(m_0)$ and $J_2(m_0, \mathcal{B}_\mathcal{N})$ are greatly differing (because in the second case the integration domain is drastically narrowed), the quantities $m_0, m_{0,\mathcal{B}_\mathcal{N}}$ and $g_0, g_{0,\mathcal{B}_\mathcal{N}}$ should also differ from each other. And this really is the case.

According to formulae (2.110),(2.111) from [23] in the general case, for *bare* quantities in the one-loop order we have

$$\begin{aligned} m_0^2 &= m_R^2 + \frac{g_R}{2}J_1(m_R) + \mathcal{O}(g_R^2) \\ g_0 &= g_R + \frac{3}{2}g_R^2 J_2(m_R) + \mathcal{O}(g_R^3). \end{aligned} \tag{49}$$

Then, considering the equalities, we can rewrite $m_{R,\mathcal{B}_\mathcal{N}} = m_R, g_{R,\mathcal{B}_\mathcal{N}} = g_R$ (49) in the one-loop order in the *measurable* picture under study as follows:

$$\begin{aligned} m_{0,\mathcal{B}_\mathcal{N}}^2 &= m_R^2 + \frac{g_R}{2}J_1(m_R, \mathcal{B}_\mathcal{N}) + \mathcal{O}(g_R^2) \\ g_{0,\mathcal{B}_\mathcal{N}} &= g_R + \frac{3}{2}g_R^2 J_2(m_R, \mathcal{B}_\mathcal{N}) + \mathcal{O}(g_R^3). \end{aligned} \tag{50}$$

BZ $\mathcal{B}_\mathcal{N}$ is a narrow low-energy (in fact central) part of the total **BZ** $\mathcal{B}$. From this it follows that the integrals $J_1(m_R, \mathcal{B}_\mathcal{N}), J_2(m_R, \mathcal{B}_\mathcal{N})$ are low-energy components of the integrals $J_1(m_R), J_2(m_R)$, respectively,and hence they are small.

As it has been noted in [24], by the lattice approach ultra-violet divergence (UVD) in QFT appear on passage to a theory in continuous space-time, i.e., for $a \to 0$. However, in this *measurable* picture we study the lattice per se rather than the continuum limit. As this takes place, UVD of a continuous theory in this case are associated with the quantities lying beyond the boundary of E_p and, in particular, beyond that of the narrowed **BZ**, i.e., $\mathcal{B}_{\mathcal{N}}$.

Because we are most interested in the experimental (renormalized) quantities of m_R, g_R which are coincident in the cases $\mathcal{B}_{\mathcal{N}}$ and $\mathcal{B}$ and defined within the energy range $E \ll E_p$, formula (50) demonstrates that *bare* quantities can be also defined at low energies $E \ll E_p$ and in terms of "narrow" **BZ** $\mathcal{B}_{\mathcal{N}}$. For the two-loop order the foregoing algorithm remains valid, excepting greater complexity of the formulae (for example formula (2.85) in [23]).

It is important that all formulae of a perturbation theory in the two-loop order in a *measurable* consideration can be derived in the same way as in the one-loop order by substitution of the short-cut Brillouin zone $\mathcal{B}_{\mathcal{N}}$ for the corresponding integrals around loop momenta over the first Brillouin zone $\mathcal{B}$.

It should be noted that the case of symmetry violation (41), i.e., $\langle\varphi(x)\rangle \neq 0$ (Section 2.2.3 in [23]) has no principal differences from our consideration. We can derive all the basic formulae in the *measurable* picture at low energies $E \ll E_p$ replacing the Brillouin zone $\mathcal{B}$ by the short-cut Brillouin zone $\mathcal{B}_{\mathcal{N}}$ in all the relevant formulae in Section 2.2.3 from [23].

Next we consider the limiting transition of this LQFT in the general case to a theory in continuous space-time, i.e., when $a \to 0$. As $a = \ell/N, N \gg 1$, we get $N \to \infty$, and from formula (11)it is inferred that full (**BZ**) $\mathcal{B} \to \infty$. It is obvious that the right and left sides of formulae (42),(49),..., where we have full (**BZ**) $\mathcal{B}$, tend to infinity. Precisely this is demonstration of UVD in canonical QFT in continuous space-time.

Since we are interested particularly in the short-cut Brillouin zone $\mathcal{B}_{\mathcal{N}}$ that is invariable, due to formulae (16) (or same (22)), the left and right sides of the corresponding formulae (43),((50)),... for $N \to \infty$ always are finite limited quantities and hence we have no UVD on passage to the continuum limit in the present consideration.

> *The principal distinction of the earlier results, e.g. [23],[24], from those obtained in this paper is the fact that in the previous works bare quantities m_0 and g_0 take infinite values on passage to the continuum limit, as is accepted by canonical QFT in continuous space-time (for example, Section 10.2 in [19]), whereas in this paper they are finite quantities obtained within the energy range $E \ll E_p$.*

4.2. Gauge-Invariante Lagrangians with the Fermions Fields

The above-mentioned results for the scalar model φ^4 are also valid for the theory of a more general type, in particular, for the Yang-Mills fields. In the lattice form, for scalar fields we can use the well-known and evaluated methods [25],for example, the Wick rotation from Minkowski space to imaginary times ((4.1) as in [25]):

$$x_0^E = i x_0^M, \\ k_0^E = -i k_0^M, \tag{51}$$

where the second line of this formula presents the Wick rotation in momentum space.

Then, for the statistical sum, the Wick rotation gives the Euclidean functional integral e^{-S_E} ((4.2) in [25])

$$e^{iS_M} \longrightarrow e^{-S_E} \tag{52}$$

the convergence of which is much better than the initial functional integral e^{iS_M}. Besides, in this case the corresponding Feynman integral of QFT, in fact, becomes the partition function of the corresponding statistical system.

In the case under study, the lattice gauge theories are most conveniently considered with the use of the approach proposed by K.Wilson [26], because it retains the gage invariance. And this is very important as, in the *measurable* form, the gauge invariance may be retained too, see Section 4.4 of [7].

Thus in the *measurable* picture we can use all the formula associated with the lattice gauge theory, specifically, the Wilson formalism [25].
We start with Euclidean action of Yang-Mills fields interaction with fermions in continuous space-time ((5.1) in [25])

$$S = \int d^4x \left[\overline{\psi}(x) \Big(\not{D} + m_f \Big) \psi(x) + \frac{1}{2} Tr \Big[F_{\mu\nu}(x) F_{\mu\nu}(x) \Big] \right] \tag{53}$$

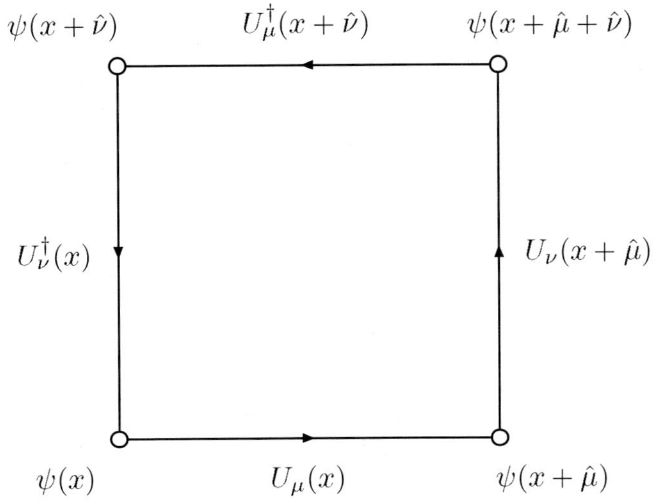

Figure 2. The plaquette.

Then its discretization according to Wilson takes the form ((5.2) given in [25])

$$\begin{aligned} S_W &= a^4 \sum_x \Big[-\frac{1}{2a} \sum_\mu \Big[\overline{\psi}(x)(r-\gamma_\mu)U_\mu(x)\psi(x+a\hat{\mu}) \\ &\quad + \overline{\psi}(x+a\hat{\mu})(r+\gamma_\mu)U_\mu^\dagger(x)\psi(x)\Big] + \overline{\psi}(x)\left(m_0 + \frac{4r}{a}\right)\psi(x)\Big] \\ &\quad + \frac{1}{g_0^2} a^4 \sum_{x,\mu\nu} \Big[N_c - \mathrm{Re}Tr[U_\mu(x)U_\nu(x+a\hat{\mu})U_\mu^\dagger(x+a\hat{\nu})U_\nu^\dagger(x)]\Big] \end{aligned} \quad (54)$$

where $x = an$, $0 < r \leq 1$ and, as usual, $\not{D} \doteq \gamma^\mu D_\mu$.

In what follows the notation is similar to that from Section 5 in [25] for the lattice spacing $a = \ell/N$, $|N| \gg 1$.

Then, according to (5.12) in [25] and by virtue of formulae (11)–(18) of this paper, in the general case for the Fourier transforms of the lattice we have

$$\psi(x) = \int_{-\frac{\pi}{a}}^{\frac{\pi}{a}} \frac{d^4p}{(2\pi)^4}\, e^{ixp}\, \psi(p) = \int\limits_{\mathcal{B}} \frac{d^4p}{(2\pi)^4}\, e^{ixp}\, \psi(p),$$

$$\overline{\psi}(x) = \int_{-\frac{\pi}{a}}^{\frac{\pi}{a}} \frac{d^4p}{(2\pi)^4}\, e^{-ixp}\, \overline{\psi}(p) = \int\limits_{\mathcal{B}} \frac{d^4p}{(2\pi)^4}\, e^{-ixp}\, \overline{\psi},$$

$$A_\mu(x) = \int_{-\frac{\pi}{a}}^{\frac{\pi}{a}} \frac{d^4k}{(2\pi)^4} e^{i(x+a\hat{\mu}/2)k}\, A_\mu(k) = \int\limits_{\mathcal{B}} \frac{d^4k}{(2\pi)^4} e^{i(x+a\hat{\mu}/2)k}\, A_\mu. \tag{55}$$

But, considering the results of Section 4 and, particularly, formulae (16),(20),(22), the formula (55) may be rewritten as

$$\psi(x) = \int\limits_{\mathcal{B}_{\mathcal{N}}} \frac{d^4p}{(2\pi)^4}\, e^{ixp}\, \psi(p),$$

$$\overline{\psi}(x) = \int\limits_{\mathcal{B}_{\mathcal{N}}} \frac{d^4p}{(2\pi)^4}\, e^{-ixp}\, \overline{\psi},$$

$$A_\mu(x) = \int\limits_{\mathcal{B}_{\mathcal{N}}} \frac{d^4k}{(2\pi)^4} e^{i(x+a\hat{\mu}/2)k}\, A_\mu. \tag{56}$$

Consequently, the Kronecker delta in position space in the general case is as follows:

$$\delta_{xy} = a^4 \int_{-\frac{\pi}{a}}^{\frac{\pi}{a}} \frac{d^4p}{(2\pi)^4} e^{i(x-y)p} = a^4 \int\limits_{\mathcal{B}} \frac{d^4p}{(2\pi)^4} e^{i(x-y)p}. \tag{57}$$

and in considered case

$$\delta_{xy} = a^4 \int\limits_{\mathcal{B}_{\mathcal{N}}} \frac{d^4p}{(2\pi)^4} e^{i(x-y)p}. \tag{58}$$

Here, the lattice spacing a is not normalized to 1 purposely, i.e., $\mathcal{B}_{\mathcal{N}}$ is given by (16).

Note that the inverse Fourier transforms in the considered *measurable* case are of the same form as in the general case and are given by formula (5.13) in

[25]

$$\begin{aligned}\psi(p) &= a^4 \sum_x e^{-ixp}\psi(x)\\ \overline{\psi}(p) &= a^4 \sum_x e^{ixp}\overline{\psi}(x)\\ A_\mu(k) &= a^4 \sum_x e^{-i(x+a\hat{\mu}/2)k} A_\mu(x)\end{aligned} \tag{59}$$

and correspondingly

$$\delta^{(4)}(p) = \frac{a^4}{(2\pi)^4}\sum_x e^{-ixp}. \tag{60}$$

However, in the general case $p \in \mathcal{B}$, $k \in \mathcal{B}$ and in the *measurable* consideration $p \in \mathcal{B}_\mathcal{N}, k \in \mathcal{B}_\mathcal{N}$.

Remark 4.1. It is convenient to use formula (22) for $\mathcal{B}_\mathcal{N}$ when the value of a is fixed. Since $\frac{\pm\pi}{N\tilde{N}a} = \frac{\pm\pi}{N\tilde{N}\ell/N} = \frac{\pm\pi}{\tilde{N}\ell}$, $\mathcal{B}_\mathcal{N}$ may be represented as a domain with the boundaries which are evidently independent of a

$$\mathcal{B}_\mathcal{N} \doteq \{p|\frac{-\pi}{\tilde{N}\ell} < p_\mu \leq \frac{\pi}{\tilde{N}\ell}, \tilde{N} \gg 1\} \doteq \mathcal{B}_{\tilde{N}}. \tag{61}$$

In this way formula (61) indicates that a ”width” (or same size) of $\mathcal{B}_{\tilde{N}}$ depends only on the number $\tilde{N}$, i.e., on EPAB. For gauge theories, in the general case we can use the same methods as in Section 4.1 with due regard for the results of Sections 2,3. Specifically, for the first-order Wilson action at gauge coupling in the general case of the quark-quark-gluon vertex in momentum space we have (formula (5.16) in [25]):

$$\begin{aligned}S_{qqg} &= -\frac{ig_0}{2}a^4 \sum_{x,\mu}\Big(\overline{\psi}(x)(r-\gamma_\mu)A_\mu(x)\psi(x+a\hat{\mu}) - \overline{\psi}(x+a\hat{\mu})(r+\gamma_\mu)A_\mu(x)\psi(x)\Big)\\ &= -\frac{ig_0}{2}a^4 \sum_{x,\mu}\int_\mathcal{B}\frac{d^4p}{(2\pi)^4}\int_\mathcal{B}\frac{d^4k}{(2\pi)^4}\int_\mathcal{B}\frac{d^4p'}{(2\pi)^4}\, e^{ix(p+k-p')}e^{iak_\mu/2}\times\\ &\times\Big(\overline{\psi}(p')(r-\gamma_\mu)A_\mu(k)\psi(p)e^{iap_\mu} - \overline{\psi}(p')e^{-iap'_\mu}(r+\gamma_\mu)A_\mu(k)\psi(p)\Big) =\\ &= \frac{ig_0}{2}\sum_\mu\int_\mathcal{B}\frac{d^4p}{(2\pi)^4}\int_\mathcal{B}\frac{d^4k}{(2\pi)^4}\int_\mathcal{B}\frac{d^4p'}{(2\pi)^4}\,(2\pi)^4\delta^{(4)}(p+k-p')e^{iak_\mu/2}\\ &\times\Big(\overline{\psi}(p')\gamma_\mu A_\mu(k)\psi(p)(e^{iap_\mu}+e^{-iap'_\mu}) + r\overline{\psi}(p')A_\mu(k)\psi(p)(-e^{iap_\mu}+e^{-iap'_\mu})\Big)\end{aligned} \tag{62}$$

$$= \frac{ig_0}{2}\sum_{\mu}\int\limits_{\mathcal{B}}\frac{d^4p}{(2\pi)^4}\int\limits_{\mathcal{B}}\frac{d^4k}{(2\pi)^4}\int\limits_{\mathcal{B}}\frac{d^4p'}{(2\pi)^4}(2\pi)^4\delta^{(4)}(p+k-p')e^{iak_\mu/2}$$
$$\times\left(\overline{\psi}(p')\gamma_\mu A_\mu(k)\psi(p)e^{iap_\mu/2}e^{-iap'_\mu/2}\cdot 2\cos\frac{a(p+p')_\mu}{2}\right.$$
$$\left.+r\overline{\psi}(p')A_\mu(k)\psi p)e^{iap_\mu/2}e^{-iap'_\mu/2}\cdot(-2i)\sin\frac{a(p+p')_\mu}{2}\right).$$

Consequently, in the considered pattern, taking into account **Remark 4.1**, we have

$$S_{qqg,N} = \frac{ig_0}{2}\sum_{\mu}\int\limits_{\mathcal{B}_{\tilde{N}}}\frac{d^4p}{(2\pi)^4}\int\limits_{\mathcal{B}_{\tilde{N}}}\frac{d^4k}{(2\pi)^4}\int\limits_{\mathcal{B}_{\tilde{N}}}\frac{d^4p'}{(2\pi)^4}(2\pi)^4\delta^{(4)}(p+k-p')e^{iak_\mu/2}$$
$$\times\left(\overline{\psi}(p')\gamma_\mu A_\mu(k)\psi(p)e^{iap_\mu/2}e^{-iap'_\mu/2}\cdot 2\cos\frac{a(p+p')_\mu}{2}\right.$$
$$\left.+r\overline{\psi}(p')A_\mu(k)\psi p)e^{iap_\mu/2}e^{-iap'_\mu/2}\cdot(-2i)\sin\frac{a(p+p')_\mu}{2}\right). \quad (63)$$

On going to the continuous limit, in the general case, i.e., for formula (62), we have

$$\lim_{a\to 0} S_{qqg} = \int_{-\infty}^{\infty}\frac{d^4p}{(2\pi)^4}\int_{-\infty}^{\infty}\frac{d^4k}{(2\pi)^4}\int_{-\infty}^{\infty}\frac{d^4p'}{(2\pi)^4}(2\pi)^4\delta^{(4)}(p+k-p')\cdot$$
$$\cdot ig_0\sum_{\mu}\overline{\psi}(p')\gamma_\mu A_\mu(k)\psi(p). \quad (64)$$

In this case we get

$$\lim_{a\to 0} S_{qqg.N} = \int\limits_{\mathcal{B}_{\tilde{N}}}\frac{d^4p}{(2\pi)^4}\int\limits_{\mathcal{B}_{\tilde{N}}}\frac{d^4k}{(2\pi)^4}\int\limits_{\mathcal{B}_{\tilde{N}}}\frac{d^4p'}{(2\pi)^4}(2\pi)^4\delta^{(4)}(p+k-p')\cdot$$
$$\cdot ig_0\sum_{\mu}\overline{\psi}(p')\gamma_\mu A_\mu(k)\psi(p). \quad (65)$$

The quantities in formulae (63) and (65) are finite as summation in the corresponding formulae is performed for the same finite domain of momenta $\mathcal{B}_{\tilde{N}}$. The difference is in the fact that the first is dependent on the lattice spacing a, whereas the second is not. But, as $a = \ell/N, N \gg 1$ is very small, the value of $S_{qqg,N}$ is close to $\lim_{a\to 0} S_{qqg.N}$. From point 2.2.2 of Section 2 it follows

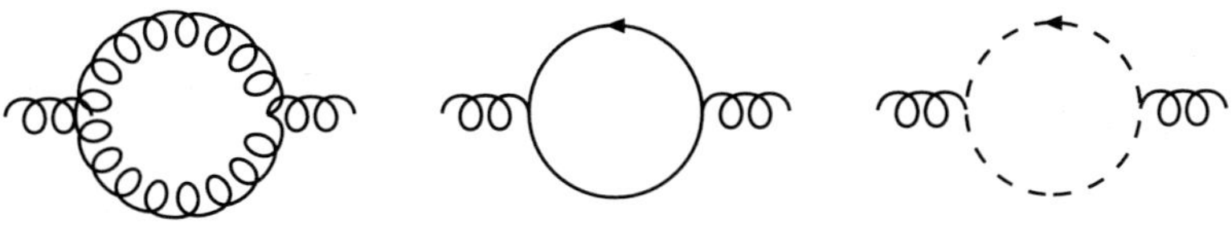

Figure 3. Loop diagrams for the self-energy of the gluon on the lattice,which have a continuum analog.

that the number $\tilde{N} \gg 1$ determines the applicability boundaries of EPAB and we always have $\tilde{N} \geq 10^2$. In the general case $\tilde{N}$ is a variable and it should be dependent on the processes under study.

All the methods considered in Sections 3,4.1 and 4.2 are also good for loop Feynman diagrams, for example, for simple loop diagrams in Figure 3. In this case, similar to formulae of Sections 4.1,4.2, the Brillouin zone $\mathcal{B}$ is contracted to $\mathcal{B}_{\tilde{N}}$,and the renormalized (experimental) quantities $m_{f,R}$ and g_R (coupling constant g_0) are coincident with $m_{f,R,\mathcal{B}_{\mathcal{N}}}$ and with $g_{R,\mathcal{B}_{\mathcal{N}}}$. In line with Section 4.1, at $a \rightarrow 0$, the initial Brillouin zone $\mathcal{B} \rightarrow \infty$. However, $\mathcal{B}_{\tilde{N}}$ remains invariant, as follows form formula (61) indicating that $\mathcal{B}_{\tilde{N}}$ is independent of a.
Then, similar to formulae (49),(50), *bare* the quantities m_0 and g_0 may be expressed in terms of the finite quantities $m_{f,R,\mathcal{B}_{\mathcal{N}}}$ and $g_{R,\mathcal{B}_{\mathcal{N}}}$, and the integrals over the finite domain $\mathcal{B}_{\tilde{N}}$. This means that they remain finite on passage to the continuous limit $a \rightarrow 0$.
From this it follows directly that in this case, similar to the scalar model φ^4, all the loop contributions in a perturbation theory may be expressed in terms of the finite quantities and integrals from the bounded (narrow,central) Brillouin zone $m_{f,R,\mathcal{B}_{\mathcal{N}}}$ that remains invariant for $a \rightarrow 0$ and leads to the finite perturbation theory both in the lattice variant and on passage to the continuous limit.
It should be noted that in the lattice consideration the number of loop Feynman diagrams may be greater. Specifically, apart from the diagrams shown in Figure 3 which have their analog in the continuous case, we can have pure lattice diagrams, in particular, to ensure the theory gauge invariance (for example, lower row in Figure 6 in [25]). Of course, all the above calculations are valid for these diagrams too.

5. Limit Transition to Continuous Theory, Equivalence Principle Applicability Boundary and Main Hypothesis

Passage from the *measurable* variant of QFT at low energies $E \ll E_p$ to the continuum limit, i.e., to the canonical quantum field theory, is possible in two ways: a) $\ell \rightarrow 0, \tau \rightarrow 0$ or b)$N_i \rightarrow \infty, N_t \rightarrow \infty$ in [6] (formula (6)). Note that, due to $\ell \approx l_p, \tau \approx t_p$ and at low energies $E \ll E_p$, $|N_i| \gg 1, |N_t| \gg 1$, the *measurable* variant of QFT is from the start close to its limit in continuous space-time.

The canonical quantum field theory in continuous space-time (QFT) [17]–[20] is a local relativistically-invariant theory considered in continuous space-time with a plane geometry, i.e with the local Minkowskian metric $\eta_{\mu\nu}(\overline{x})$. And this assumption is valid for all the energy range. Still, it is quite clear that the quantum processes associated with QFT (particle collisions, decay,...) can introduce perturbations into the space-time geometry, varying its curvature. But as QFT is a local theory, a strong Equivalence Principle (EP) [27] enables one, in a sufficiently small region $\mathcal{V}_r$ of the fixed point, to consider space-time as a flat space in this case too. Consequently, we naturally think about the applicability boundary of this principle.

As noted in [8],[9], the natural applicability boundary of EP is associated with the Planck energies $E \approx E_p$ or same Planck scales $l \approx l_p$. Currently, there is no doubt that at very high energies (on the order of Planck energies $E \approx E_p$), i.e., on Planck scales, $l \approx l_p$ quantum fluctuations of any metric $g_{\mu\nu}(\overline{x})$ are so high that in this case the geometry determined by $g_{\mu\nu}(\overline{x})$ is replaced by the "geometry" following from *quantum foam* that is defined by great quantum fluctuations of $g_{\mu\nu}(\overline{x})$, i.e., by the characteristic spatial sizes of the quantum-gravitational region (for example, [28]–[33]). The above-mentioned geometry is drastically differing from the locally smooth geometry of continuous space-time and EP in it is no longer valid [34]–[41]. Actually, the *quantum foam* is not geometry in a common sense as locally it is determined by a set of different metrics, each of which is taken into consideration with its statistical weight [31]. From this it follows that the region $\mathcal{V}_{\overline{r},\overline{t}}$ with the characteristic spatial size $\overline{r} \approx l_p$ (and hence with the temporal size $\overline{t} \approx t_p$) EP is no longer valid. According to the present-day knowledge, *quantum foam* is an aggregation of bubbles (cells) of Planck's size, each of which has some micro(tiny)-manifold with a given

geometry (for example, micro black hole (mbh), micro wormhole, etc.)
Main Hypothesis. [9],[10].
It is assumed that in the general case EP, and consequently, QFT is valid for the locally smooth space-time only if all the energies E of the particles are satisfied the necessary condition

$$E \ll E_p, \tag{66}$$

It should be noted that, provided the hypothesis is true, it should be true no matter whether QFT is studied in the perturbative or non-perturbative mode, because the local geometry of space-time is not dependent on this fact.
In [10] *Main Hypothesis* has been proven in the assumption that space-time foam consists of micro black holes (**mbh**), with the event horizon radius $r \approx l_p$ and mass $m \approx m_p$. Note that a model for the *quantum foam* comprising micro black holes (mbh) is most probable from the viewpoint of a correct passage to low energies [33],[38],[39].

Remark 5.1. Why in canonical QFT it is so important never forget about the fact that space-time has a flat geometry, or the same possesses the Minkowskian metric $\eta_{\mu\nu}(\overline{x})$? Simply, in the contrary case we should refuse from some fruitful methods and from the results obtained by these methods in canonical QFT, in particular from Wick rotation [20]. In fact, in this case the time variable is replaced by $t \mapsto it \doteq t_E$, and the Minkowskian metric $\eta_{\mu\nu}(\overline{x})$ is replaced by the four-dimensional Euclidean metric

$$ds^2 = dt_E^2 + dx^2 + dy^2 + dz^2. \tag{67}$$

Clearly, such replacement is possible only in the case when from the start space-time (locally) has a flat geometry, i. e. possesses the Minkowskian metric $\eta_{\mu\nu}(\overline{x})$. This is another argument supporting the key role of the EP applicability boundary. Otherwise, when we go beyond this boundary, Wick rotation becomes invalid. Naturally, some other methods of canonical QFT will lose their force too.

6. The Strong Equivalence Principle, Black Holes Thermodynamics and QFT

It is supposed that a large (i. e, classical) four-dimensional Schwarzschild black hole is existent with the metric

$$ds^2 = \left(1 - \frac{2MG}{r}\right) dt^2 - \left(1 - \frac{2M}{r}\right)^{-1} dr^2 - r^2 d\Omega^2, \tag{68}$$

where M is the mass of this black hole, and the Schwarzschild horizon radius r_{BH} is defined by

$$r_{BH} = 2MG. \tag{69}$$

As shown in [42],[43], EP is violated for an observer distant from the black hole event horizon. Considering our objective, it seems expedient to give in brief the main results from [42],[43].

In view of the Unruh effect, an accelerating observer does detect thermal radiation (so-called Unruh radiation) with the Unruh temperature given by [44]

$$T_{Unruh} = \frac{\hbar a}{2\pi}, \tag{70}$$

where $a = |\mathbf{a}|$ is a corresponding acceleration.

When an observer is at the *fixed* distance, $r > r_{BH}$, from a Schwarzschild black hole of mass M and event horizon radius $r_{BH} = 2GM$, then, due to the existence of Hawking radiation [45], the observer will measure radiation with thermal spectrum and a temperature given by formula [42]

$$T_{H,r} = \frac{\hbar}{8\pi GM\sqrt{1 - \frac{r_{BH}}{r}}}, \tag{71}$$

where $r > r_{BH}$. In the foregoing formulae and in what follows, we use the normalization $c = k_B = 1$.

To detect Unruh radiation, one can use an Unruh-Dewitt detector [46]–[48]. This detector is an idealized point particle with the internal energy levels labeled by the energy E, coupled via the monopole interaction with a scalar field ϕ. The detector moves along the world line described by the functions $x^\mu(\tau)$, where τ is the detector's proper time. The detector-field interaction is described by the interaction Lagrangian $cm(\tau)\phi[x(\tau)]$, where c is a small coupling constant and

m is the detector's monopole moment operator. Suppose the field ϕ is in the vacuum state $|0_M>$, where $|0_M>$ is the "Minkowski vacuum" [46].

For a general trajectory, the detector will not remain in its ground state E_0, but will undergo a transition to an excited state $E > E_0$, while the field will make a transition to an excited state $|\psi>$. For sufficiently small c the amplitude for this transition may be given by first order perturbation theory as [46]

$$ic\langle E,\psi|\int_{-\infty}^{\infty} m(\tau)\phi[x(\tau)]d\tau|0_M,E_0\rangle \tag{72}$$

The transition amplitude (72), in virtue of the time evolution of $m(\tau)$

$$m(\tau) = e^{H_0\tau}m(0)e^{-H_0\tau},$$

may be transformed (factorized) as

$$ic\langle E|m(0)|E_0\rangle\psi|\int_{-\infty}^{\infty} e^{(E-E_0)\tau}\langle\psi|\phi(x)|0_M\rangle d\tau. \tag{73}$$

Here $H_0|E\rangle = E|E\rangle$ [46].

Then, squaring the modulus (73) and summing over E and the complete set ψ, one can calculate the transition probability for all possible E and ψ:

$$c^2\sum_E |\langle E|m(0)|E_0\rangle|^2\mathcal{F}(E-E_0), \tag{74}$$

where $\mathcal{F}(E)$ is the detector *response function*

$$\mathcal{F}(E) = \int_{-\infty}^{\infty} d\tau \int_{-\infty}^{\infty} d\tau' e^{-E(\tau-\tau')}G^+(x(\tau),x(\tau')). \tag{75}$$

The detector response function $\mathcal{F}(E)$ is independent of the its details, is determined by the positive frequency Wightman Green function G^+ and represents the bath of "particles" effectively experienced by the detector as a result of its motion [46].

Within the scope of the Unruh effect, a number of the above-mentioned particles (Rindler particles) in the Minkowski vacuum comes to $|0_M>$ –this is a thermal Planck spectrum of Rindler particles with temperature given by formula (70).

In [42] it is shown that an observer, positioned at the fixed distance $r > r_{BH}$ from the above-mentioned black holes and measuring Hawking temperature with the value $T_{H,r}$, experiences the local acceleration

$$a_{BH,r} = \frac{1}{\sqrt{1-\frac{r_{BH}}{r}}}\left(\frac{r_{BH}}{2r^2}\right). \tag{76}$$

Another observer in the Einstein elevator, moving with acceleration through Minkowskian space-time, will measure the same acceleration toward the floor of the elevator, thermal radiation with the Unruh temperature given by formula (70). As shown in [42], a_{BH} is coincident with the quantity a from the formula in (70). Then substituting the acceleration $a = a_{BH}$ from formula (76) into formula (70), we can obtain a formula for $T_{Unruh,r}$ in this case [43]:

$$T_{Unruh,r} = \frac{\hbar}{2\pi\sqrt{1-\frac{r_{BH}}{r}}}\left(\frac{r_{BH}}{2r^2}\right). \tag{77}$$

If EP is valid, the quantities $T_{Unruh,r}$ from formula (77) and $T_{H,r}$ in (71) should be coincident for $r > r_{BH}$ to a high degree of accuracy. However, we see that this is not true. In [43], e.g. for $r = 4GM = 2r_{BH}$, we have $T_{H,r} = 4T_{Unruh,r}$.

So, far from the event horizon, EP is not the case. Moreover, violation of EP is the greater the farther it is from the black hole event horizon. Indeed, for an observer at the distance $r > r_{BH}$ we can write $r = \alpha(r)GM = \frac{1}{2}\alpha(r)r_{BH}, \alpha(r) > 2$. Then

$$T_{Unruh,r} = \frac{\hbar}{2\pi\alpha^2(r)GM\sqrt{1-\frac{2}{\alpha(r)}}}. \tag{78}$$

In this way $T_H/T_{Unruh,r} = \alpha^2(r)/4$. And the ratio is the greater, the higher $\alpha(r)$, i.e., the farther from horizon the observer is. Next, for compactness, we denote $T_{Unruh,r}$ in terms of $T_{U,r}$. Of course, in this case we bear in mind only an observer at a sufficiently great but finite distance from a black hole, i.e., only when a gravitational field is thought significant and must be taken into consideration. So, in the general case $r_{BH} < r \ll \infty$, whereas in the case of a distant observer we have

$$r_{BH} \ll r \ll \infty. \tag{79}$$

Obviously, this case of violated EP is not directly associated with the *Main Hypothesis* concerning the boundaries of EP validity (formula (66)) from the previous section, because in [42],[43] consideration is given to a large black hole with the event horizon radius r_{BH} much greater than Planck length $r_{BH} \gg l_p$ at sufficiently low energies.

Really, the resulting distribution of the particles emitted by a black hole has the form (last formula on p.122 in [49])

$$n_E = \Gamma_{gb}[exp(\frac{E}{T_H}) - 1]^{-1}, \tag{80}$$

where n_E is the number of particles with the energy E and $\Gamma_{gb} < 1$ is the so-called *greybody factor*. As the black hole mass M is large, the temperature T_H is low, and then from the last formula it follows that arbitrary large values of n_E will be given only by particles with the energies E close to a small value of T_H.

The principal result from the remarkable papers [42],[43] may be summarized as follows:

Comment 6.1. In any point of space-time that is in a field of a large classical Schwarzschild black hole, and in the cases when this field must be taken into consideration, it is impossible to remove this field in the vicinity of the point even locally, i.e., to consider space-time as flat.

Comment 6.2. It is important to refine some formulations from [42],[43]. Specifically, if $r \to r_{BH}$, then $T_{H,r} \to \infty, T_{Unruh,r} \to \infty$ in formulae (71) and (77), respectively. Note that for $r \to r_{BH}$ these temperatures become infinite $T_{H,r} = \infty, T_{Unruh,r} = \infty$. Based on this fact, in [42],[43] it is inferred "that the equivalence principle is restored on the horizon". But this statement is not correct. Restoration of EP is not following from the fact that the above temperatures take infinite values. We can only state that temperature on the BH horizon and in its vicinity cannot be the parameter detecting a deviation from EP. In the opposite case one can arrive at violation: on the black hole event horizon, where a gravitational field is very large in value, EP holds, whereas far from the event horizon, where a gravitational field is much weaker, this principle is violated.

The question arises: what changes in the foregoing pattern are associated with the substitution of **mbh** for a large four-dimensional Schwarzschild black hole? Note that **mbh** has Planck's parameters, i.e., we have

$$r_{mbh} \propto l_p, m_{mbh} \propto m_p. \tag{81}$$

However,it is supposed that **mbh** is not an *extremal black hole*, i.e., **mbh** with minimal values of $r_{mbh} = r_{min}, m_{mbh} = r_{min}$, as the latter have no Hawking radiation [50]. Without loss of generality, it is assumed that the abovementioned **mbh** is sufficiently close to *extremal black hole*:

$$r_{mbh} \approx r_{min}, m_{mbh} \approx m_{min}, \tag{82}$$

but is not coincident with it $r_{mbh} \neq r_{min}, m_{mbh} \neq m_{min}$.

According to the present-day knowledge, a semi-classical region starts between 5 and 20 times the Planck scale [51]. Then, despite the fact that this **mbh** is itself in the quantum gravity region, the scales of $r \gg r_{mbh}$ determine the region, where a semi-classical approximation is valid.

Let us return to high energy physics and to the subject of the previous section. One of the preferable models for space-time foam is the model based on the assumption that its unit cells are **mbh**, with radius and mass on the order of the Planck (for example, [33],[38], [39]. Of great importance for **mbh** are the quantum-gravitational effects and the corresponding quantum corrections of black hole thermodynamics at Planck scale (for example, [52]).

Then, in line with formula (20) in [52], we have minimal values for radius and mass of a black hole

$$r_{min} = \sqrt{\frac{e}{2}}\alpha^{'} l_p, \quad m_{min} \doteq m_0 = \frac{\alpha^{'}\sqrt{e}}{2\sqrt{2}} m_p, \tag{83}$$

where the number $\alpha^{'}$ is on the order of 1, and in [52] we take the normalization $\hbar = c = k_B = 1$ in which $l_p = m_p^{-1} = T_p^{-1} = \sqrt{G}$.

From (83) it directly follows that the formula for the event horizon radius $r = 2MG$, valid for large classical black holes, will be valid in the case when we include the quantum-gravitational effects for **mbh** because $r_{min} = 2m_0G$. Such a black hole of a minimal size is associated with a maximal temperature (formula (24) in [52]):

$$T_H^{max} = \frac{T_p}{2\pi\sqrt{2}\alpha^{'}}. \tag{84}$$

A black hole satisfying the formulae (83),(84) is termed as *minimal* (or *Planck*).

For the energies E somewhat lower than the Planck energies (i.e., $E \ll E_p$) involved in the condition (66), a semiclassical approximation is valid. This means that, on substitution of **mbh** with the mass m_{mbh} for a large (classical) black hole with the mass M, in the case under study the results, substantiated when an observer uses the standard Unruh-Dewitt detector in radiation measurement for coupled to a massless scalar field [46],[48], are valid with the corresponding quantum corrections [53].

Let us revert to the formulae from [42],[43]. In particular, to formula (76) for the real acceleration measured by an observer who is positioned at the fixed distance $r \gg r_{BH}$ in the Schwarzschild space-time, given by (formula (14) in [42], formula (9.170) in [54])

$$a_{BH,r} = a_S = \frac{\sqrt{\nabla_\mu V \nabla^\mu V}}{V} = \frac{MG}{r^2\sqrt{1-2MG/r}} = \frac{MG}{r^2 V}, \tag{85}$$

where it is supposed that a static observer at the radius r moves along orbits of the time-like Killing vector $K = \partial_t$ and $V = \sqrt{-K_\mu K^\mu} = \sqrt{1-2MG/r}$ is the red-shift factor for the Schwarzschild space-time (p.413 in [54]).

How changes formula (85) on going to **mbh**? It is clear that the condition $r \gg r_{BH}$ is replaced by the condition $r \gg r_{mbh}$ (corresponding to the condition $E \ll E_p$ and semiclassical approximation),M is replaced by m_{mbh},the red-shift factor V should be replaced by V_q, where V_q is the quantum deformation of V with regard to quantum corrections in the field **mbh**. Then, for **mbh**, formula (85) is of the form

$$a_{S,q} = \frac{m_{mbh}G}{r^2 V_q}, \tag{86}$$

where $a_{S,q}$ is the real acceleration with regard to the quantum corrections measured by a distant observer in the field **mbh**.

Clearly, formula (71) for $T_{H,r}$, due to formulae for the red-shift factor V, in the general case may be given as

$$T_{H,r} = \frac{\hbar}{8\pi GMV}\,. \tag{87}$$

Then its quantum analog, i.e., the corresponding formula for temperature in the field **mbh**, for $r \gg r_{mbh}$ is as follows:

$$T_{H,r,q} = \frac{\hbar}{8\pi G m_{mbh} V_q}. \tag{88}$$

In virtue of formula (85), formula (77) takes the form

$$T_{U,r} = \frac{\hbar}{2\pi\sqrt{1-\frac{r_{BH}}{r}}}\left(\frac{r_{BH}}{2r^2}\right) = \frac{\hbar}{2\pi V}\left(\frac{r_{BH}}{2r^2}\right). \tag{89}$$

For $r > r_{BH}$ and due to formula (78), we have

$$T_{U,r} = \frac{\hbar}{2\pi\alpha^2(r)GM\sqrt{1-\frac{2}{\alpha(r)}}} = \frac{\hbar}{2\pi\alpha^2(r)GMV}. \tag{90}$$

As indicated above, for $r \gg r_{BH}$ we have $\alpha(r) \gg 1$.

What are the changes on going to **mbh**?

Considering the case $r \gg r_{mbh}$ and semiclassical picture, we again come to $\alpha(r) \gg 1$, whereas formula (90) is replaced by formula

$$T_{U,r,q} = \frac{\hbar}{2\pi\alpha^2(r)Gm_{mbh}V_q}. \tag{91}$$

Proceeding from the above, we have

$$\frac{T_{H,r,q}}{T_{U,r,q}} = \frac{T_{H,r}}{T_{U,r}} = \frac{\alpha^2(r)}{4} \tag{92}$$

Formula (92) points to the fact that, within the scope of a semiclassical approximation, relations of a black hole temperature to the Unruh temperature for a distant observer in the case of a large (classical) black hole and **mbh** are coincident because these quantities are dependent on the same factors:
first, on $1/MV$ and, second, on $1/m_{mbh}V_q$.

Note that in this consideration there s no need to have an explicit formula for V_q as this quantity is not involved in the key expression (92). Specifically, to derive an explicit expression for V_q, we can use the results from [55] on quantum deformation of the Schwarzschild solution due to spherically symmetric quantum fluctuations of the metric and the matter fields. In this case the Schwarzschild singularity at $r = 0$ is shifted to the finite radius $r_{min} \approx r_{mbh} \propto l_p$, where the scalar curvature is finite. In this way the results from [55] correlate well with the results from [52].

Quantum corrections at Planck scales were obtained in [52] proceeding from validity of the Generalized Uncertainty Principle (GUP)[11]–[14]. But the results presented in this work are independent of this aspect. Actually, during studies of black hole thermodynamics at Planck scales with the use of other methods [56],[57] (differing from those in [52]),in particular, Loop Quantum Gravity (LQG)[57], the obtained results were similar to [52]. Because of this, for **mbh** with all the thermodynamic characteristics (mass,radius,temperature,...) on the order of the corresponding Planck quantities, all the calculations in this section are valid. As noted above, far from horizon of **mbh**, i.e., at the energies $E \ll E_p$ (66), the results from [42],[43] remain valid in this case as well.

Next, similar to [33], we assume that in every cell of space-time foam a micro black hole (**mbh**) with a typical gravitational radius of $r_{min} \propto l_p$ may be present. Then, in according with the results in [42],[43] and in virtue of the formula (92) we come to violation of the strong EP for distance r, satisfying the condition

$$l_p \ll r \ll \infty, \tag{93}$$

that is equivalent to $\widetilde{E}_r \ll E_p$ for the energies $\widetilde{E}_r$ associated with the scale of r.

In the last formula it is implicitly (and purely conditionally) assumed that a minimum length is equal to l_p and to r_{min}. But, as noted above, in the general case we have $r_{min} \propto l_p$, i.e., the order is similar to that of l_p. Specifically, in [59] by natural assumptions it has been demonstrated that the minimum length may be twice and more as great as the Planck length. It is obvious that all the above calculations and derivations of the present work are independent of the specific value of r_{min}.

In this way, if the quantum foam structure is determined by **mbh**, the applicability of QFT is limited to the energies $E < \widetilde{E}_r \ll E_p$ and the formula (66) is the case. This supports the it Main Hypothesis from Section 2 within the assumption concerning the quantum foam structure made in this section.

Comment 6.3. In the case of **mbh** Comment 6.2 is absolutely clear. In fact, at a horizon of **mbh**, i.e., for $T_{H,r,q} = T_{Unruh,r,q} = \infty$ similar to large black holes but , naturally, without any restoration of EP as the domain $r = r_{mbh} \approx r_{min} \propto l_p$ is the region of Planck energies or of quantum foam, where EP in its canonical formulation becomes invalid. It is obvious that at the event horizon $r = r_{mbh}$ of **mbh** and in its vicinity a gravitational field becomes very strong due to quantum effects and nothing could destroy it.

7. Final Comments and Conclusion

7.1. Based on the above results, all the energies E we can classify into 3 groups [10]:

a) low energies $0 < E \leq \widetilde{E} \ll E_p$ – energies, for which the Strong Equivalence Principle is valid in virtue of formula (66), and hence this energy interval sets the QFT applicability boundaries.
a1) Since $\widetilde{E} \ll E_p$, it is natural to assume that $\widetilde{E} \approx 10^{-N} E_p$, where $N \geq 2$. Obtaining of more accurate estimates for N is a separate problem;

b) intermediate energies $\widetilde{E} < E < E_p$ – energies, for which the Strong Equivalence Principle and, consequently QFT, becomes invalid but the corresponding scales are greater than the Planck. It can be assumed that QFT in this energy range will be a theory in a gravitational field that could not be destroyed even locally. In the case under study it is assumed that this field is created by **mbh**. Impossibility of destroying this field even locally is associated with large quantum corrections for the corresponding quantities which should be taken into consideration at these energies [52], [56], [57].
Let us call the energy scale $\widetilde{E} < E < E_p$ as *prequantum gravity phase*;

c) high (essentially maximal) energies $E \approx E_p$ or $E > E_p$. This interval is the region of quantum gravity energies.

Next note that, as all the experimentally involved energies E are low, they satisfy condition a) or b). Specifically, for LHC, maximal energies are $\approx 10TeV = 10^4 GeV$, that is by 15 orders of magnitude lower than the Planck energy $\approx 10^{19} GeV$. Moreover, the characteristic energy scales of all fundamental interactions also satisfy condition a). Indeed, in the case of strong interactions this scale is $\Lambda_{QCD} \sim 200 MeV$; for electroweak interactions this scale is determined by the vacuum average of a Higgs boson and equals $v \approx 246 GeV$; finally, the scale of the (Grand Unification Theory (GUT)) M_{GUT} lies in the range of $\sim 10^{14} GeV - -10^{16} GeV$.
It should be noted, however, that on validity of assumption a1) the energy scale M_{GUT} lies within the applicability region of the energy group a) and hence of QFT. Provided the EP applicability boundaries are lying at considerably lower energies, a study of GUT necessitates a theory with (even locally) unremovable

curvature.
At the same time, it is clear that the requirement of the Lorentz-invariant QFT, due to the action of Lorentz boost (or same hyperbolic rotations), results in however high momenta and energies. But it has been demonstrated that unlimited growth of the momenta and energies is impossible because in this case we fall within the energy region, where the conventional quantum field theory [17]–[20] is invalid.
Note that at the present time there are experimental indications that Lorentz-invariance is violated in QFT on passage to higher energies (for example, [58]). Besides, one should note important recent works associated with EP applicability boundaries and violation in nuclei and atoms at low energies (for example [60]). Proceeding from the above, the requirement for Lorentz-invariance and EP is possible only within the scope of the condition (66).

7.2. Proceeding from the above results, it is inferred that the well-known QFT [17]–[20], from the start, is a ultraviolet-finite theory with the natural cutoff parameter $l_{\widetilde{E}} \propto \hbar/\widetilde{E}$. Note that the quantum-gravitational parameter $r_{mbh} \propto l_p$ is beyond the applicability limits of QFT [10].

7.3. In the present approach it is of interest to study the problem of *asymptotic safety* introduced by Steven Weinberg in [61]. This problem was addressed in [10] as suggested by the Reviewers. We use the definition of this notion given in ([62],p.67): "A theory is said to be asymptotically safe if all essential coupling parameters g_j (these are the ones that are invariant under field redefinitions) approach, for energies $k \rightarrow \infty$, a fixed point where at least one of them does not vanish." If initially it has been assumed that r_{min} is considered within the scope of (GUP) [11]–[14], this definition necessitates certain refinements. In particular, if (GUP) supposes the existence of a maximum momentum p_{max}, as in [13] or (Section V in [14]), it is clear that the condition $k \rightarrow \infty$ can not be fulfilled. So, the condition $k \rightarrow \infty$ should be replaced by the condition $k \rightarrow p_{max}$. Most often, it is assumed that the momentum p_{max} is on the order of the Planck momentum, i.e., we have $p_{max} \propto p_{pl}$. However, in the most general case the quantity p_{max} may be even trans-Planck.
When p_{max} is inexistent (i.e., $p_{max} = \infty$), still by this approach of *asymptotic safety* the problem should be reformulated in accordance with the fact (shown above) that, beginning with the energies $E, E > \widetilde{E}$, a theory must be considered as QFT in curved space-time in a field created by **mbh**. In his further works the

author is planning to study this problem within the scope of this approach in greater detail.

7.4. It is absolutely clear that, though the material in Section 4 and the results in it are associated with the canonical QFT in continuous space-time, they are still valid for the discrete variant of a theory given in Section 2. In [8], [9] a quantum theory is considered only for the energies satisfying the conditions $E \ll E_p$ and $E \approx E_p$ rather than for a complete scale of energies. In [10] and in the present paper this gap is eliminated.

7.5. Returning to Section 4, we should note an important difference between a semiclassical consideration in [42], [43] and results in [10] or in this paper. In the first case a **UDW** detector may be without limit close to the classical BH event horizon, because in this case the semiclassical approximation is still valid. But in the case under study a **UDW** detector should be sufficiently distant from the **mbh** event horizon, otherwise it can fall into the quantum-gravity region and become meaningless.

7.6. The subject-matter of this work is associated with such a problem as experimental detection of a quantum foam. At the present time, the relevant results are still inconsistent [63] due to inadequate capacity of the data available and insufficient precision of the experiments. (The situation resembles that in chronology of the detection of gravitational waves relict including.) Nevertheless, an active search in this direction is in progress [64], [65].

7.7. Finally, it is important to note that Albert Einstein formulated his strong Equivalence Principle only for classical objects against a classical space-time background. This is explained by the fact that in the period, when General Relativity was created (1915,1916) and this principle was formulated, a quantum theory was at the stage of conception and its main postulates were unknown. Because of this, formulation of the principal result from [42], [43] is not quite exact. Strong EP, in its initial formulation, is still valid and the principal result from [42], [43] should be as follows:
generalization of the strong Equivalence Principle for the case of a semiclassical approximation in a black hole gravitational field becomes invalid.

Thus, the main problem of QFT applicability for discrete and continuous consideration in perturbative or nonperturbative mode is associated with definition of the scale $\widetilde{E}$ *in point 7.1.*

Conflict of Interests

The author declares that there is no conflict of interests regarding the publication of this work.

References

[1] Shalyt-Margolin, A.E. Minimal Length and the Existence of Some Infinitesimal Quantities in Quantum Theory and Gravity. *Adv. High Energy Phys.,* **2014** (2014), 8. http://dx.doi.org/10.1155/2014/195157.

[2] Shalyt-Margolin, Alexander. Minimal Length, Measurability and Gravity. *Entropy* **2016**, *18(3)*, 80. http://dx.doi.org/10.3390/e18030080.

[3] Shalyt-Margolin,Alexander. The Uncertainty Principle and Minimal Length at All Energy Scales: Some Implications. Chapter 2 in *Advances in Quantum Field Theory Research* **2017**, 33–73. Nova Science Publishers 2017.

[4] Shalyt-Margolin, Alexander E. Minimal Length, Minimal Inverse Temperature, Measurability and Black Hole. *Electronic Journal of Theoretical Physics* **2018**, *14(37)*, 35-54.

[5] Shalyt-Margolin, Alexander. Measurability Notion in Quantum Theory, Gravity and Thermodynamics. Basic Facts and Implications. Chapter 8 in *Horisons in World Physics* **2017**, *292*, 199 - 244. Nova Science Publishers 2017.

[6] Shalyt-Margolin, Alexander. Minimal Quantities and Measurability. Gravity in Measurable Format and Natural Transition to High Energies. *Nonlinear Phenomena in Complex Systems* **2018**, *21(2)*, 138 - 163.

[7] Shalyt-Margolin A., The Equivalence Principle Applicability Boundaries, QFT in Flat Space and Measurability I. Free Quantum Fields. *Nonlinear Phenomena in Complex Systems* **2019**, *22(2)*, 135 - 150.

[8] Shalyt-Margolin A., QFT in Flat Space and Measurability II. Perturbation Theory for a Scalar Field Model. *Nonlinear Phenomena in Complex Systems* **2020**, *23(1)*, 33 - 53.

[9] Shalyt-Margolin A., The Equivalence Principle Applicability Boundaries, Measurability, and UVD in QFT. *Int. J. Nonlinear Phenomena in Complex Systems* **24(1)**, 38 - 55, (2021).

[10] Shalyt-Margolin A., The Quantum Field Theory Boundaries Applicability and Black Holes Thermodynamics. *Int. J. Theor. Phys.* (2021). https://doi.org/10.1007/s10773-021-04804-1.

[11] Adler R. J. and Santiago D. I., On gravity and the uncertainty principle. *Mod. Phys. Lett. A*, *14*, 1371–1378 (1999).

[12] Kempf A., Mangano G., and Mann R.B., Hilbert space representation of the minimal length uncertainty relation. *Phys. Rev. D.*, **52**, 1108–1118 (1995).

[13] Nozari Kourosh, Etemadi Amir, Minimal length, maximal momentum and Hilbert space representation of quantum mechanics, *Phys. Rev. D.*, **85**, 104029 (2012).

[14] Tawfik Abdel Nasser, Diab Abdel Magied. Generalized Uncertainty Principle: Approaches and Applications. *Int. J. Mod. Phys. D*, **23**, 1430025 (2014).

[15] Heisenberg W., Uber den anschaulichen Inhalt der quantentheoretischen Kinematik und Mechanik [About the clear content of the quantum theoretical kinematics and mechanics]. *Z. Phys.*, **43** (1927), 172–198. (In German) http://dx.doi.org/10.1007/bf01397280.

[16] Messiah, A. *Quantum Mechanics*; North Holland Publishing Company: Amsterdam, The Netherlands, Volumes 1, 2.

[17] Rayder, Lewis H. *Quantum Field Theory*, University of Kent and Canterbuty.

[18] Cheng Ta-Pei, Li Ling-Fong, *Gauge Theory of elementary particle physics*, Oxford Scince Publications, 548 pages.

[19] Peskin M. E., Schroeder D.V., *An Introduction to Quantum Field Theory*, Addison-Wesley Publishing Company, 1995.

[20] Weinberg Steven, *The Quantum Theory of Fields*, Vol.1, 2. Cambridge University Press, 1995.

[21] Faddeev, L., Mathematical View on Evolution of Physics. *Priroda* **1989**, *5*, 11.

[22] Smit Jan, *Introduction Quantum Fields Lattice* (Cambridge Lecture Notes in Physics), 2002.

[23] Montvay, Istvan and Munster, Gernot (1994). *Quantum Fields on a Lattice.* Cambridge University Press, Cambridge.

[24] Hernandez Pilar, *Introduction to Lattice Field Theory*, Universidad de Valencia and IFIC, 2010.

[25] Capitani Stefano, Lattice Perturbation Theory. *Phys. Rept* **2003**, *382*, 113-302.

[26] Wilson K.G. (1974). Confinement Of Quarks, *Phys. Rev.* **D10**, 2445.

[27] Weinberg, S. *Gravitation and Cosmology. Principles and Applications of General Theory of Relativity*. 1972.

[28] Wheeler, J. A. "Geons". *Phys. Rev.* **1955**, *97*, 511.

[29] Wheeler, J. A. *Geometrodynamics* (Academic Press, New York and London, 1962).

[30] Misner, C. W., Thorne, K. S. and Wheeler, J. A. *Gravitation* Freeman, San Francisco, (1973).

[31] Hawking, S. W., Space-time foam. *Nuclear Phys. B* **1978**, *114*, 349.

[32] Ng Y. J., Selected topics in Planck-scale physics, *Mod. Phys. Lett. A.*, vol.18, pp. 1073-1098, 2003.

[33] Scardigli, Fabio, Black Hole Entropy: a spacetime foam approach, *Class. Quant.Grav.*, **14** (1997), 1781–1793.

[34] Garattini, Remo, A Spacetime Foam approach to the cosmological constant and entropy. *Int. J. Mod. Phys. D.* **4** (2002) 635–652.

[35] Garattini, Remo, A Spacetime Foam Approach to the Schwarzschild-de Sitter Entropy. *Entropy* **2** (2000) 26–38.

[36] Garattini, Remo, Entropy and the cosmological constant: a spacetime-foam approach. *Nucl. Phys. Proc. Suppl.* **88** (2000) 297–300.

[37] Garattini, Remo, Entropy from the foam. *Phys. Lett. B* **1999**, *459*, 461–467.

[38] Scardigli, Fabio. Generalized Uncertainty Principle in Quantum Gravity from Micro-Black Hole Gedanken Experiment. *Phys. Lett. B.* **1999**, *452*, 39–44.

[39] Scardigli, Fabio. Gravity coupling from micro-black holes. *Nucl. Phys. Proc. Suppl.* **2000**, *88*, 291–294.

[40] Garay, Luis J. Thermal properties of spacetime foam. *Phys. Rev. D* **1998**, *58*, 124015.

[41] Garay, Luis J. Spacetime foam as a quantum thermal bath. *Phys. Rev. Lett.* **1998**, *80*, 2508–2511.

[42] Singleton, D. and Wilburn, S., Hawking Radiation, Unruh Radiation, and the Equivalence Principle, *Phys. Rev. Lett.*, **107**, 081102 (2011); *Phys. Rev. Lett.*, **108**, 049002 (2012).

[43] Singleton, Douglas and Wilburn, Steve, Global versus local–Mach principle versus the equivalence principle, *Int. J. Mod. Phys. D.*, **25**,1644009 (2016).

[44] Unruh, W. G., Notes on black-hole evaporation, *Phys. Rev. D.*, **14**, 870 (1976).

[45] Hawking, S. W., Particle Creation by Black Holes, *Comm. Math. Phys.* **43**,199 (1975).

[46] Birrell, N. D. and Davies, P. C. W., *Quantum fields in curved space*, Cambridge University Press, Cambridge (1982).

[47] De Witt, B. S., Quantum gravity: new synthesis, in *General Relativity*, Eds. S. Hawking, W. Israel, Cambridge University Press, 1979.

[48] Akhmedov, E. T. and Singleton D., On the relation between Unruh and Sokolov–Ternov effects, *Int. J. Mod. Phys. A*, **22**, 4797 (2007).

[49] Mukhanov, V. F. and Winitzki, S., *Introduction to Quantum Fields in Classical Backgrounds*, Lecture notes (2004).

[50] Carroll, S. M., Johnson, M. C. and Randall L. Extremal limits and black hole entropy. *Journal of High Energy Physics.* **11**, 109 (2009). doi:10.1088/1126-6708/2009/11/109.

[51] Meade, P. and Randall, L., Black holes and quantum gravity at the LHC. *Journal of High Energy Physics.* **05**, 003 (2008). doi:10.1088/1126-6708/2008/05/003

[52] Nouicer, K., Quantum-corrected black hole thermodynamics to all orders in the Planck length, *Phys. Lett B*, **646**, 63 (2007).

[53] Kiefer, Claus, The semiclassical approximation to quantum gravity, in *Canonical Gravity: From Classical to Quantum*, pp.170–212, edited by J. Ehlers and H. Friedrich, Springer, Berlin (1994).

[54] Carroll, S. M., *Spacetime and Geometry. An Introduction to General Relativity*, Addison-Wesley, San Francisco (2014).

[55] Kazakov, D. I., Solodukhin, S. N., On Quantum Deformation of the Schwarzschild Solution. *Nucl.Phys. B.*, **429**, 153-176 (1994).

[56] Jamil, Mubasher, Darabi, Farhad, Quantum corrections for a black hole in an asymptotically safe gravity with higher derivatives, *Int. J. Theor. Phys.* **51**, 1007-1012 (2012).

[57] Gorji, M. A., Nozari Kourosh, Vakili B., Polymeric Quantization and Black Hole Thermodynamics, *Phys. Lett. B.*, **735**, 1007-1012 (2014).

[58] Kostelecky, V.A., Russell, N., Data tables for Lorentz and CPT violation. *Rev. Mod. Phys. 83(1)*, 11–31 (2011).

[59] Rong-Jia, Yang and Shuang, Nan Zhang, Modified clock inequalities and modified black hole lifetime, *Phys. Rev. D.*, **79**, 124005 (2009), arXiv:0906.2108 [gr-qc].

[60] Flambaum, V. V., Enhanced violation of the Lorentz invariance and Einstein's equivalence principle in nuclei and atoms. *Phys. Rev. Lett.* **2016**, *117*, 072501.

[61] Weinberg, Steven, Ultraviolet divergences in quantum theories of gravitation, in *General Relativity: An Einstein centenary survey*, pp. 790-831, edited by S. W. Hawking and W. Israel, Cambridge University Press, (1979).

[62] Kiefer, Claus, *Quantum Gravity. Third Edition*, Oxford University Press, United Kingdom (2012).

[63] *Integral challenges physics beyond Einstein* / Space Science / Our Activities. ESA, Noordwijk, Netherlands, July 1, 2011.

[64] Moyer, Michael (17 January 2012). "Is Space Digital?". *Scientific American*. Retrieved 3 February 2013.

[65] Cowen, Ron (22 November 2012). "Single photon could detect quantum-scale black holes". *Nature News*. Retrieved 3 February 2013.

In: Quantum Field Theory and Applications
Editor: N. Palerma
ISBN: 978-1-68507-930-7

Chapter 3

An Alternative Discrete Quantum Field Theory

Stan Gudder*
Department of Mathematics
University of Denver, Denver, Colorado, USA

Abstract

We present a version of quantum field theory based on discrete energy-momentum. Beginning with annihilation and creation operators, we construct interaction quantum number operators. These are employed to define discrete quantum fields. We then discuss scattering operators and their properties. The theory is illustrated using examples of fermion and boson fields.

Keywords: quantum field theory, discrete energy-momentum

1. Introduction

It is well known that traditional quantum field theory is not mathematically precise. The main reason for this is that some of its concepts are not rigorously defined and this results in a multitude of singularities and divergences [5, 6]. One way of overcoming these problems is to assume that space-time and energy-momentum are discrete [1, 2]. In this way, space-time and energy-momentum

*Corresponding Author's Email: sgudder@du.edu.

differences are not allowed to approach zero so singularities do not occur [3, 4]. This idea has been resisted because space-time discreteness has not been observed and it violates Lorentz invariance [5, 6]. In this work we present an alternative proposal in which only energy-momentum are assumed to be discrete. We then define energy-momentum instead of space-time quantum fields. We can trace this back to the origins of quantum mechanics when Max Planck postulated the existence of discrete energy packets. Of course, we go further than this by also assuming discrete momenta components.

In this work we only present the beginning outlines of a theory. If it shows promise, then more details can be pursued later. Our main purpose is to define the scattering operator and discuss some of its properties. Section 2 presents the basic definitions and Sections 3 and 4 illustrate the theory using examples of fermion and boson fields.

2. Basic Definitions

Let $Z = \{0, \pm 1, \pm 2, \ldots\}$ be the set of integers and $Z^+ = \{0, 1, 2, \ldots\}$ be the set of nonnegative integers. We call $\mathcal{S} = Z^+ \times Z^3$ the *discrete energy-momentum space.* If $p = (p_0, p_1, p_2, p_3) \in \mathcal{S}$ we write $p = (p_0, \underline{p})$ where $p_0 \in Z^+$ represents the total energy and $\underline{p} \in Z^3$ the momentum of a particle or anti-particle. Ordinarily, we would write $p = (\omega_0 p_0, \omega_1 \underline{p})$ where ω_0, ω_1 are small elementary units of energy and momentum, but for simplicity we set $\omega_0 = \omega_1 = 1$ by assuming these scales are unity. We equip $\mathcal{S}$ with the Minkowski distance

$$\|p\|_4^2 = p_0^2 - \|\underline{p}\|_3^2 = p_0^2 - p_1^2 - p_2^2 - p_3^2$$

We assume that $\|p\|_4^2 \geq 0$ and call $m = \pm \|p\|_4$ the *mass* of the particle (antiparticle) when $m \geq 0 (m < 0)$. We conclude that Einstein's energy formula $p_0 = \sqrt{m^2 + \|\underline{p}\|_3^2}$ holds. This theory predicts that there are a countable number of admissible particle (antiparticle) masses. If m is a particle (antiparticle) mass, we call

$$\Gamma_m = \{p \in \mathcal{S} \colon \|p_4\| = m\}$$

the *mass hyperboloid.* We define the *geometric speed* of $p \in \mathcal{S}$ by

$$v = \frac{\|\underline{p}\|_3}{p_0} = \frac{\sqrt{p_0^2 - m^2}}{p_0} = \sqrt{1 - \frac{m^2}{p_0^2}}$$

when $p_0 \neq 0$. Of course, $0 \leq v \leq 1$ and $v = 1$ if and only if $m = 0$. Moreover, $v = 0$ if and only if $\|\underline{p}\| = 0$.

Particle-antiparticles are distinguished according to their mass and spin. If p has integer spin, then p is a *boson* and if p has half-integer spin, then p is a *fermion*. We only consider theories in which there are a finite number of particle types. For example, suppose we have two types of fermions $p_1, p_2, \ldots, q_1, q_2, \ldots$, and two types of bosons $r_1, r_2, \ldots, s_1, s_2, \ldots$. Then we have masses m_i, $i = 1, 2, 3, 4$ and $p_k \in \Gamma_{m_1}$, $q_k \in \Gamma_{m_2}$, $r_k \in \Gamma_{m_3}$, $s_k \in \Gamma_{m_4}$, $k = 1, 2, \ldots$. The theory is described by a Hilbert space H with orthonormal basis $|0\rangle$, ψ_{nmuv} where

$$\psi_{nmuv} = |p_1 \cdots p_n q_1 \cdots q_n r_1 \cdots r_u s_1 \cdots s_v\rangle \tag{1}$$

The vector $|0\rangle$ is the *vacuum state* which represents the condition in which no particles are present. The vector ψ_{nmuv} is the state in which there are n fermions of type p, m fermions of type q, u bosons of type r and v bosons of type s. It is assumed that ψ_{nmuv} is symmetric under the interchange of bosons and particles of different types, while it is antisymmetric under the interchange of fermions of the same type [5, 6]. For example,

$$|p_1 p_2 q_1 q_2 r_1 r_2\rangle = |p_1 q_1 p_2 q_2 r_2 r_1\rangle$$

while

$$|p_1 p_2 q_1 q_2 r_1 r_2\rangle = -|p_2 p_1 q_1 q_2 r_1 r_2\rangle = -|p_1 p_2 q_2 q_1 r_1 r_2\rangle$$

It follows that any basis vector has at most one fermion with a particular mass, spin and energy-momentum p. Thus, for ψ_{nmuv} we have $p_i \neq p_j$ and $q_i \neq q_j$ for $i \neq j$.

The fundamental operators of the theory are the *annihilation* and *creation* operators a_p, a_p^* on $H, p \in \Gamma_m$ where $a_p|0\rangle = 0$, $0 \in H$ is the zero vector and $a_p^*|0\rangle = |p\rangle$. If p_j is a fermion and ψ has the form (1), then

$$a_{p_j}\psi_{nmuv} = \pm|p_1 \cdots p_{j-1} p_{j+1} \cdots p_n q_1 \cdots\rangle$$

where the $-$ sign applies if j is even and the $+$ sign applies if j is odd. Moreover, if $p \in \Gamma_{m_1}$ and $p \neq p_j$ for any j, then

$$a_p^*\psi_{nmuv} = |pp_1 \cdots p_n q_1 \cdots\rangle$$

If r is a boson and r appears n times we use the notation

$$|p_1 \cdots p_n r^n r_1 \cdots r_u \cdots\rangle = |p_1 \cdots p_n rr \cdots r_1 \cdots r_u \cdots\rangle$$

where there are n r's on the right-hand side. In this case, we know that

$$a_r|p_1 \cdots p_n r^n r_1 \cdots r_u \cdots\rangle = \sqrt{n}\,\left|p_1 \cdots p_n r^{n-1} r_1 \cdots r_u \cdots\right\rangle$$

and

$$a_r^*|p_1 \cdots p_n r^n r_1 \cdots r_u \cdots\rangle = \sqrt{n+1}\,\left|p_1 \cdots p_n r^{n+1} r_1 \cdots r_u \cdots\right\rangle$$

In general, if p does not appear in ψ, then $a_p\psi = 0$. We define the particle number operator $A_p, p \in \Gamma_m$ by $A_p = a_p^* a_p$. Then A_p is a self-adjoint operator on H and for a state $\psi \in H$, $A_p\psi = n\psi$ where n is the number of times that p appears in ψ. Of course, if p is a fermion, then n is 0 or 1. We call the operator G on H given by

$$G = \sum_{p\in\Gamma_m} \frac{1}{p_0} A_p$$

a *free quantum field.* The terms $1/p_0$ ensure that the sum converges to a self-adjoint operator on H. If p and q are particle types, we define the *interaction number operator* $A_{pq}, p \in \Gamma_{m_1}, q \in \Gamma_{m_2}$ by

$$A_{pq} = \tfrac{1}{2}\left[a_p^* a_q + a_q^* a_p\right]$$

Then A_{pq} is a self-adjoint operator on H that we shall study in the next two sections. Notice that if we have $p = q$, then $A_{pp} = A_p$.

We now define energy-momentum quantum fields and scattering operators. In order to avoid complicated indices, we shall consider a theory in which there are four particle types with states of form (1). The reader can then easily visualize how the general situation develops. For particles p, q, r, s of the four types we define the *energy-momentum field* F on H by

$$F = \frac{1}{6}\left[\sum_{p,q} \frac{1}{p_0+q_0} A_{pq} + \sum_{p,r} \frac{1}{p_0+r_0} A_{pr} + \sum_{p,s} \frac{1}{p_0+s_0} A_{ps} \right.$$
$$\left. + \sum_{q,r} \frac{1}{q_0+r_0} A_{qr} + \sum_{q,s} \frac{1}{q_0+s_0} A_{qs} + \sum_{r,s} \frac{1}{r_0+s_0} A_{rs}\right] \quad (2)$$

where $p \in \Gamma_{m_1}$, $q \in \Gamma_{m_2}$, $r \in \Gamma_{m_3}$, $s \in \Gamma_{m_4}$. As with G, it is not hard to show that (2) defines F as a self-adjoint operator on H. In particular, if there are two types p, q of particles, then

$$F = \sum \left\{ \frac{1}{p_0 + q_0} A_{pq} \colon p \in \Gamma_{m_1}, q \in F_{m_2} \right\}$$

and if $p = q$, then $F = G$ the free quantum field. We define the *scattering operator* to be the unitary operator on H given by $S = e^{i\omega F}$ where $\omega \geq 0$ is a constant called the *relaxation time*. We consider ω to be a constant of nature depending on the particle types. If $\psi \in H$ is an initial state for the system, then we interpret $S\psi$ to be the final scattered state. We define the transition probability $|\langle \phi, S\psi \rangle|^2$ to be the probability that the final state is ϕ. Computing $S\psi$ exactly is usually quite difficult and approximations of the form

$$S = e^{i\omega F} = \sum_{n=0}^{\infty} \frac{1}{n!}(i\omega F)^n \approx I + i\omega F - \frac{\omega^2}{2} F^2 - \frac{i\omega^3}{3!} F^3 + \frac{\omega^4}{4!} F^4 \qquad (3)$$

are made. We call (3) the fifth approximation and higher order approximations give more exact values.

In the case of a free field G, there is no scattering. The reason for this is that every basis vector is an eigenvector for G. For example if $p_1 \neq p_2$ then

$$G|p_1 p_2\rangle = \sum_p \tfrac{1}{p_0} A_p |p_1 p_2\rangle = \left(\tfrac{1}{p_{1,0}} + \tfrac{1}{p_{2,0}} \right) |p_1 p_2\rangle$$

Hence, if $\psi \in H$ is a basis vector, there exists $\lambda \in \mathbb{R}$ such that $G\psi = \lambda\psi$. It follows that

$$S\psi = e^{i\omega G}\psi = e^{i\omega\lambda}\psi$$

Hence, the transition probability $|\langle \psi, S\psi \rangle|^2 = 1$ so there is no scattering.

We define the *energy-momentum operators* E_i, $i = 0, 1, 2, 3$ by

$$E_i = \sum_{p \in \Gamma_m} p_i A_p = \sum_{p \in \Gamma_m} p_i a_p^* a_p$$

It is clear that A_p, $p \in \Gamma_m$ and the free quantum field $G = \sum_{p \in \Gamma_m} \frac{1}{p_0} A_p$ commute with E_i, $i = 0, 1, 2, 3$ and this illustrates conservation of energy-momentum.

We define the *interaction energy-momentum* operators in the natural way by addition. For example, if there are two particle types, say p and q-types, then

$$E_i = \sum_{p,q} (p_i a_p^* a_p + q_i a_q^* a_q), \quad i = 0, 1, 2, 3$$

Now E_i do not commute with the interaction operators A_{pg} and

$$F = \sum_{p,q} \frac{1}{p_0 + q_0} A_{pq}$$

Indeed,

$$E_0 A_{p_1 q_1} |p_1\rangle = \tfrac{1}{2}\, E_0 |q_1\rangle = \tfrac{1}{2}\, q_{1,0} |q_1\rangle$$

and

$$A_{p_1 q_1} E_0 |p_1\rangle = p_{1,0} A_{p_1 q_1} |p_1\rangle = \tfrac{1}{2} p_{1,0} |q_1\rangle$$

These do not agree if $q_{1,0} \neq p_{1,0}$. Moreover,

$$E_0 F |p_1\rangle = \tfrac{1}{2} \sum_q \frac{1}{p_{1,0} + q_0} E_0 |q\rangle = \tfrac{1}{2} \sum_q \frac{q_0}{p_{1,0} + q_0} |q\rangle$$

and

$$F E_0 |p_1\rangle = p_{1,0} F |p_1\rangle = \tfrac{1}{2} p_{1,0} \sum_q \frac{1}{p_{1,0} + q_0} |q\rangle$$

Again, these do not agree, in general. We conclude that energy-momentum need not be conserved when there is an interaction. This says that particles can absorb or lose energy-momentum to the interaction field.

Also, observe that although the particle number operators commute, the interaction number operators do not, in general. Indeed, if p_1, q_1, q_2 are fermions we have that

$$A_{p_1 q_2} A_{p_1 q_1} |p_1 q_1\rangle = 0$$

but on the other hand

$$A_{p_1 q_1} A_{p_1 q_2} |p_1 q_1\rangle = \tfrac{1}{2} A_{p_1 q_1} |q_2 q_1\rangle = -\tfrac{1}{4} |p_1 q_2\rangle$$

If these particles are bosons we conclude that

$$A_{p_1q_2}A_{p_1q_1}|p_1q_1^2\rangle = \tfrac{1}{2}A_{p_1q_2}\left(\sqrt{3}|q_1^3\rangle + 2|p_1^2q_1\rangle\right)$$
$$= A_{p_1q_2}|p_1^2q_1\rangle = \frac{\sqrt{2}}{2}|p_1q_1q_2\rangle$$

but on the other hand

$$A_{p_1q_1}A_{p_1q_2}|p_1q_1^2\rangle = \tfrac{1}{2}A_{p_1q_1}|q_1^2q_2\rangle = \frac{\sqrt{2}}{4}|p_1q_1q_2\rangle$$

However, we have the following result which is useful for later computations.

Lemma 1. *If $p \neq p_1$ and $q \neq q_1$, then*

$$A_{pq}A_{p_1q_1} = A_{p_1q_1}A_{pq} \tag{4}$$

Proof. We first assume that the four particles are bosons. It is easy to see that (4) holds if and only if it holds on states of the form $|p^r p_1^s q^t q_1^u\rangle$. We then have

$$\begin{aligned}
&A_{pq}A_{p_1q_1}|p^r p_1^s q^t q_1^u\rangle \\
&= \frac{1}{2}A_{pq}\left[\sqrt{u(s+1)}\,|p^r p_1^{s+1}q^t q_1^{u-1}\rangle + \sqrt{s(u+1)}\,|p^r p_1^{s-1}q^t q_1^{u+1}\rangle\right] \\
&= \tfrac{1}{4}\Big[\sqrt{u(s+1)t(r+1)}\,|p^{r+1}p_1^{s+1}q^{t-1}q_1^{u-1}\rangle \\
&\quad +\sqrt{u(s+1)r(t+1)}\,|p^{r-1}p_1^{s+1}q^{t+1}q_1^{u-1}\rangle \\
&\quad +\sqrt{s(u+1)t(r+1)}\,|p^{r+1}p_1^{s-1}q^{t-1}q_1^{u+1}\rangle \\
&\quad +\sqrt{s(u-1)r(t+1)}\,|p^{r-1}p_1^{s-1}q^{t+1}q_1^{u+1}\rangle\Big]
\end{aligned}$$

Moreover, we have that

$$\begin{aligned}
&A_{p_1q_1}A_{pq}|p^r p_1^s q^t q_1^u\rangle \\
&= \tfrac{1}{2}A_{p_1q_1}\left[\sqrt{t(r+1)}\,|p^{r+1}p_1^s q^{t-1}q_1^u\rangle + \sqrt{r(t+1)}\,|p^{r-1}p_1^s q^{t+1}q_1^u\rangle\right] \\
&= \tfrac{1}{4}\Big[\sqrt{t(r+1)u(s+1)}\,|p^{r+1}p_1^{s+1}q^{t-1}q_1^{u-1}\rangle \\
&\quad +\sqrt{t(r+1)s(u+1)}\,|p^{r+1}p_1^{s-1}q^{t-1}q_1^{u+1}\rangle \\
&\quad +\sqrt{r(t+1)u(s+1)}\,|p^{r-1}p_1^{s+1}q^{t+1}q_1^{u-1}\rangle \\
&\quad +\sqrt{r(t+1)s(u+1)}\,|p^{r-1}p_1^{s-1}q^{t+1}q_1^{u+1}\rangle\Big]
\end{aligned}$$

The cases in which some of the particles are fermions follow by replacing the corresponding superscripts with ones. □

3. Fermions

This section illustrates the theory outlined in the previous section with an example involving two fermion types that we call type-p with mass m_1 and type-q with mass m_2, where $m_1 \neq m_2$. The Hilbert space H has orthonormal basis $|0\rangle, |p_1 \cdots p_n q_1 \cdots q_m\rangle$ where $p_i \neq p_j$, $q_i \neq q_j$, $i \neq j$. We have the conditions $|p_i q_j\rangle = |q_j p_i\rangle$, $|p_i p_j\rangle = -|p_j p_i\rangle$, $|q_i q_j\rangle = -|q_j q_i\rangle$ as well as their higher order generalizations. Notice that all annihilation and creation operators commute except

$$a_p a_p^* + a_p^* a_p = a_q a_q^* + a_q^* a_q = I$$

It is easy to check that A_{pq} has only three eigenvalues $0, \pm 1/2$, each having infinite-dimensional eigenspaces. For example, if $p \neq p_i$, $i = 1, \ldots, n$, $q \neq q_j$, $j = 1, \ldots m$, then $A_{pq}|p_1 \cdots p_n q_1 \cdots q_m\rangle = 0$. Moreover, it follows from $A_{pq}|p\rangle = \frac{1}{2}|q\rangle$ and $A_{pq}|q\rangle = \frac{1}{2}|p\rangle$ that if

$$\psi_\pm = |pp_1 \cdots p_n q_1 \cdots q_m\rangle \pm |qp_1 \cdots p_n q_1 \cdots q_m\rangle$$

then

$$A_{pq}(\psi_+ + \psi_-) = \tfrac{1}{2}(\psi_+ + \psi_-)$$
$$A_{pq}(\psi_+ - \psi_-) = -\tfrac{1}{2}(\psi_+ - \psi_-)$$

The energy-momentum field is given by

$$F = \sum \left\{ \tfrac{1}{p_0 + q_0} A_{pq} \colon p \in \Gamma_{m_1}, q \in \Gamma_{m_2} \right\} = \sum_{p,q} \frac{1}{p_0 + q_0} A_{pq}$$

In particular, $F|0\rangle = 0$,

$$F|p_1\rangle = \sum_{p,q} \frac{1}{p_0 + q_0} A_{pq}|p_1\rangle = \sum_q \frac{1}{p_{1,0} + q_0} A_{p_1 q}|p_1\rangle = \sum_q \frac{1}{2(p_{1,0} + q_0)} |q\rangle \quad (5)$$

and similarly $F|q_1\rangle = \sum\limits_p \frac{1}{2(p_0 + q_{1,0})} |p\rangle$. In general, we have that

$$F|p_1 \cdots p_n q_1 \cdots q_m\rangle = \sum_{p,q} \frac{1}{p_0 + q_0} A_{pq}|p_1 \cdots p_n q_1 \cdots q_m\rangle$$
$$= \sum_{q \neq q_1, \ldots, q_m} \frac{1}{2(p_{1,0} + q_0)} |p_2 \cdots p_n q q_1 \cdots q_m\rangle$$

$$
\begin{aligned}
&- \sum_{q\neq q_1,\ldots,q_m} \frac{1}{2(p_{2,0}+q_0)}|p_1p_3\cdots p_nqq_1\cdots q_m\rangle \\
&+\cdots+(-1)^{n+1} \sum_{q\neq q_1,\ldots,q_m} \frac{1}{2(p_{n,0}+q_0)}|p_1\cdots p_{n-1}qq_1\cdots q_m\rangle \\
&+ \sum_{p\neq p_1,\ldots,p_n} \frac{1}{2(p_0+q_{1,0})}|pp_1\cdots p_nq_2\cdots q_m\rangle \\
&- \sum_{p\neq p_1,\ldots,p_n} \frac{1}{2(p_0+q_{2,0})}|pp_1\ldots p_nq_1q_3\ldots q_m\rangle \\
&+\cdots+(-1)^{m+1} \sum_{p\neq p_1,\ldots,p_n} \frac{1}{2(p_0+q_{m,0})}|pp_1\cdots p_nq_1\cdots q_{m-1}\rangle
\end{aligned}
$$

Suppose we are interested in the decay of a particle p_1. We would then want the scattered state $S|p_1\rangle = e^{i\omega F}|p_1\rangle$. From (3), to a fifth order of approximation we have that

$$
S|p_1\rangle \approx |p_1\rangle + i\omega F|p_1\rangle - \frac{\omega^2}{2}F^2|p_1\rangle - i\frac{\omega^3}{3!}F_3|p_1\rangle + \frac{\omega^4}{4!}F^4|p_1\rangle
$$

We have computed $F|p_1\rangle$ in (5). For $F^2|p_1\rangle$ we obtain

$$
\begin{aligned}
F^2|p_1\rangle &= \frac{1}{2}\sum_{1}\frac{1}{p_{1,0}+q_0}F(q) = \frac{1}{4}\sum_{q}\frac{1}{p_{1,0}+q_0}\sum_{p'}\frac{1}{p'_0+q_0}|p'\rangle \\
&= \frac{1}{4}\sum_{q,p'}\frac{1}{(p_{1,0}+q_0)(p'_0+q_0)}|p'\rangle
\end{aligned}
$$

Continuing we obtain

$$
\begin{aligned}
F^3|p_1\rangle &= \frac{1}{4}\sum_{q_1p'}\frac{1}{(p_{1,0}+q_0)(p'_0+q_0)}F|p'\rangle \\
&= \frac{1}{8}\sum_{q,p'q'}\left[(p_{1,0}+q_0)(p'_0+q_0)(p'_0+q'_0)\right]^{-1}|q'\rangle \\
F^4|p_1\rangle &= \frac{1}{16}\sum_{q,p',q',p''}\left[(p_{1.0}+q_0)(p'_0+q_0)(p'_0+q'_0)(p''_0+q'_0)\right]^{-1}|p''\rangle
\end{aligned}
$$

For a state $|p_2\rangle$ we have that

$$\begin{aligned}\langle p_2|S|p_1\rangle &\approx \langle p_2 \mid p_1\rangle - \frac{\omega^2}{2}\langle p_1|F^2|p_1\rangle + \frac{\omega^2}{4!}\langle p_2|F^4|p_1\rangle \\ &= \langle p_2 \mid p_1\rangle - \frac{\omega^2}{4\bullet 2}\sum_q [(p_{1,0}+q_0)(p_{2,0}+q_0)]^{-1} \\ &\quad + \frac{\omega^4}{16\bullet 4!}\sum_{q,p',q'}\left[(p_{1,0}+q_0)(p'_0+q_0)(p'_0+q'_0)(p_{2,0}+q'_0)\right]^{-1}\end{aligned}$$

Of course, if $p_2 \neq p_1$ then $\langle p_2 \mid p_1\rangle = 0$ and if $p_2 = p_1$ then $\langle p_2 \mid p_1\rangle = 1$. The transition probability can be computed from $|\langle p_2|S|p_1\rangle|^2$.

As another example, suppose we want the scattering probabilities for an initial state $|p_1p_2\rangle$. To the third order of approximation we have that

$$S \approx I + i\omega F - \frac{1}{2}\omega^2 F^2$$

We then compute

$$F|p_1p_2\rangle \sum_{p,q}\frac{1}{p_0+q_0}A_{pq}|p_1p_2\rangle = \frac{1}{2}\sum_q \frac{1}{p_{1,0}+q_0}|qp_2\rangle - \frac{1}{2}\sum_q \frac{1}{p_{2,0}+p_0}|qp_1\rangle$$

and

$$\begin{aligned}F^2|p_1p_2\rangle &= \frac{1}{2}\sum_q \frac{1}{p_{1,0}+q_0}F|qp_2\rangle - \frac{1}{2}\sum_q \frac{1}{p_{2,0}+q_0}F|qp_1\rangle \\ &= \frac{1}{2}\sum_q \frac{1}{p_{1,0}+q_0}\sum_{p',q'}\frac{1}{p'_0+q'_0}A_{p'q'}|qp_2\rangle \\ &\quad - \frac{1}{2}\sum_q \frac{1}{p_{2,0}+q_0}\sum_{p',q'}\frac{1}{p'_0+q'_0}A_{p'q'}|qp_1\rangle \\ &= \frac{1}{4}\sum_q \frac{1}{p_{1,0}+q_0}\sum_{q'\neq q}\frac{1}{p_{2,0}+q'_0}|qq'\rangle \\ &\quad + \frac{1}{4}\sum_q \frac{1}{p_{1,0}+q_0}\sum_{p'\neq p_2}\frac{1}{p'_0+q_0}|p'p_2\rangle \\ &\quad - \frac{1}{4}\sum_q \frac{1}{p_{2,0}+q_0}\sum_{q'\neq q}\frac{1}{p_{1,0}+q'_0}|q'q\rangle \\ &\quad - \frac{1}{4}\sum_q \frac{1}{p_{2,0}+q_0}\sum_{p'\neq p_1}\frac{1}{p'_0+q_0}|p'p_1\rangle\end{aligned}$$

We then obtain

$$\langle p_1p_2|S|p_1p_2\rangle \approx 1-\frac{\omega^2}{2}\langle p_1p_2|F^2|p_1p_2\rangle = 1-\frac{\omega^2}{8}\sum_q\left[\frac{1}{(p_{1,0}+q_0)^2}+\frac{1}{(p_{2,0}+q_0}\right]$$

Moreover,

$$\langle p_3p_2|S|p_1p_2\rangle \approx 1-\frac{\omega^2}{8}\sum_q\frac{1}{(p_{1,0}+q_0)(p_{3,0}+q_0)}$$

From these we compute the transition probabilities.

4. Bosons

The situation for bosons is more complicated than for fermions because more than one identical particles can appear. We saw for fermions that the operators A_{pq} had only the eigenvalues $0, \pm 1/2$ with corresponding infinite-dimensional eigenspaces. We now conjecture that for bosons, the operators A_{pq} have the eigenvalues $0, \pm 1/2, \pm 1, \pm 3/2, \pm 2, \ldots$ again, with infinite-dimensional eigenspaces. Suppose our system contains two bosons p and q of different types (and other particles that we are not concerned about). We call the subspace S_n^{pq} generated by $\left\{\left|p^iq^j\right\rangle : i+j=n\right\}$ the *n-subspace* of H for A_{pq}. If an eigenvalue λ of A_{pq} has a corresponding eigenvector in S_n^{pq}, we call λ an *n-eigenvalue* of A_{pq}. Of course, A_{pq} has eigenvalue 0 because $A_{pq}|0\rangle = 0$. Moreover, we can append different particles, $p_1, q_1, \ldots$ to get infinitely many other eigenvectors $|p_1q_1\ldots\rangle$ of A_{pq} corresponding to the eigenvalue 0. We now present a sequence of examples that make our conjecture very plausible.

Example 1. *Let S_1^{pq} be the 1-subspace for A_{pq} generated by $\{|p\rangle, |q\rangle\}$. Since*

$$A_{pq}\left(|p\rangle+|q\rangle\right)=\tfrac{1}{2}\left(|p\rangle+|q\rangle\right)$$

and

$$A_{pq}\left(|p\rangle-|q\rangle\right)=\tfrac{1}{2}\left(|p\rangle-|q\rangle\right)$$

We conclude that $\pm 1/2$ are 1-eigenvalues of A_{pq}. As before, we can append different particles to obtain infinitely many eigenvectors of A_{pq} corresponding to eigenvalues $\pm 1/2$. This comment also applies to our later example. □

Example 2. *Let S_2^{pq} be the 2-subspace for A_{pq} generated by $\{|p^2\rangle, |q^2\rangle, |pq\rangle\}$. Since $A_{pq}\left(|p^2\rangle - |q^2\rangle\right) = 0$ and*

$$\begin{aligned}
A_{pq}\left(|p^2\rangle + |q^2\rangle + \sqrt{2}\,|pq\rangle\right) &= \tfrac{1}{2}\left(2\sqrt{2}\,|pq\rangle + 2|q^2\rangle + 2|p^2\rangle\right)\\
&= |p^2\rangle + |q^2\rangle + \sqrt{2}\,|pq\rangle\\
A_{pq}\left(|p^2\rangle + |q^2\rangle - \sqrt{2}\,|pq\rangle\right) &= \tfrac{1}{2}\left(2\sqrt{2}\,|pq\rangle - 2|q^2\rangle - 2|p^2\rangle\right)\\
&= -\left(|p^2\rangle + |q^2\rangle - \sqrt{2}\,|pq\rangle\right)
\end{aligned}$$

we conclude that $0, \pm 1$ are 2-eigenvalues of A_{pq}. □

Example 3. *Let S_3^{pq} be the 3-subspace for A_{pq} generated by $\{|p^3\rangle, |q^3\rangle, |p^2q\rangle, |pq^2\rangle\}$. Since*

$$\begin{aligned}
&A_{pq}\left(|p^3\rangle + |q^3\rangle + \sqrt{3}\,|p^2q\rangle + \sqrt{3}\,|pq^2\rangle\right)\\
&= \frac{\sqrt{3}}{2}\left(|p^2q\rangle + |pq^2\rangle\right) + \frac{\sqrt{3}}{2}\left(2|pq^2\rangle + \sqrt{3}\,|p^3\rangle\right) + \frac{\sqrt{3}}{2}\left(\sqrt{3}\,|q^3\rangle + 2|p^2q\rangle\right)\\
&= \frac{3}{2}\left(|p^3\rangle + |q^3\rangle + \sqrt{3}\,|p^2q\rangle + \sqrt{3}\,|pq^2\rangle\right)\\
&A_{pq}\left(|p^3\rangle - |q^3\rangle - \sqrt{3}\,|p^2q\rangle + \sqrt{3}\,|pq^2\rangle\right)\\
&= \frac{\sqrt{3}}{2}\left(|p^2q\rangle - |pq^2\rangle\right) - \frac{\sqrt{3}}{2}\left(2|pq^2\rangle + \sqrt{3}\,|p^3\rangle\right) + \frac{\sqrt{3}}{2}\left(\sqrt{3}\,|q^3\rangle + 2|p^2q\rangle\right)\\
&= -\frac{3}{2}\left(|p^3\rangle - |q^3\rangle - \sqrt{3}\,|p^2q\rangle + \sqrt{3}\,|pq^2\rangle\right)\\
&A_{pq}\left(|p^3\rangle - |q^3\rangle + \frac{\sqrt{3}}{3}\,|p^2q\rangle - \frac{\sqrt{3}}{3}\,|pq^2\rangle\right)\\
&= \frac{\sqrt{3}}{2}\left(|p^2q\rangle - |pq^2\rangle\right) + \frac{\sqrt{3}}{6}\left(2|pq^2\rangle + \sqrt{3}\,|p^3\rangle\right) - \frac{\sqrt{3}}{6}\left(\sqrt{3}|q^3\rangle + 2|p^2q\rangle\right)\\
&= \frac{1}{2}\left(|p^3\rangle - |q^3\rangle + \frac{\sqrt{3}}{3}\,|p^2q\rangle - \frac{\sqrt{3}}{3}\,|pq^2\rangle\right)\\
&A_{pq}\left(|p^3\rangle + |q^3\rangle - \frac{\sqrt{3}}{3}\,|p^2q\rangle - \frac{\sqrt{3}}{3}\,|pq^2\rangle\right)\\
&= \frac{\sqrt{3}}{2}\left(|p^2q\rangle + |pq^2\rangle\right) - \frac{\sqrt{3}}{6}\left(2|pq^2\rangle + \sqrt{3}\,|p^3\rangle\right) - \frac{\sqrt{3}}{6}\left(\sqrt{3}|q^3\rangle + 2|p^2q\rangle\right)\\
&= -\frac{1}{2}\left(|p^3\rangle + |q^3\rangle - \frac{\sqrt{3}}{3}\,|p^2q\rangle - \frac{\sqrt{3}}{3}\,|pq^2\rangle\right)
\end{aligned}$$

we conclude that $\pm\frac{3}{2}, \pm\frac{1}{2}$ are 3-eigenvalues of A_{pq}. □

Example 4. *Let S_4^{pq} be the 4-subspace for A_{pq} generated by*

$$\{|p^4\rangle, |q^4\rangle, |p^3q\rangle, |pq^3\rangle, |p^2q^4\rangle\}$$

Instead of proceeding as we did in the previous examples, we display the matrix representing A_{pq} in the above basis and the eigenpairs for this matrix. We then have that

$$A_{pq} = \begin{bmatrix} 0 & 0 & 2 & 0 & 0 \\ 0 & 0 & 0 & 2 & 0 \\ 2 & 0 & 0 & 0 & \sqrt{6} \\ 0 & 2 & 0 & 0 & \sqrt{6} \\ 0 & 0 & \sqrt{6} & \sqrt{6} & 0 \end{bmatrix}$$

Also, $\pm 2, \pm 1, 0$ are 4-eigenvalues of A_{pq} with the corresponding (unnormalized) eigenvectors given by:

$$\begin{bmatrix} 1 \\ 1 \\ 2 \\ 2 \\ \sqrt{6} \end{bmatrix}, \quad \begin{bmatrix} 1 \\ 1 \\ -2 \\ -2 \\ \sqrt{6} \end{bmatrix}, \quad \begin{bmatrix} 1 \\ -1 \\ 1 \\ -1 \\ 0 \end{bmatrix}, \quad \begin{bmatrix} 1 \\ -1 \\ -1 \\ 1 \\ 0 \end{bmatrix}, \quad \begin{bmatrix} 1 \\ 1 \\ 0 \\ 0 \\ -\sqrt{6}/3 \end{bmatrix} \qquad \square$$

Example 5. *Let S_5^{pq} be the 5-subspace for A_{pq} generated by*

$$\{|p^5\rangle, |q^5\rangle, |p^4q\rangle, |pq^4\rangle, |p^3q^2\rangle, |p^2q^3\rangle\}$$

We then obtain

$$A_{pq} = \begin{bmatrix} 0 & 0 & \sqrt{5} & 0 & 0 & 0 \\ 0 & 0 & 0 & \sqrt{5} & 0 & 0 \\ \sqrt{5} & 0 & 0 & 0 & 2\sqrt{2} & 0 \\ 0 & \sqrt{5} & 0 & 0 & 0 & 2\sqrt{2} \\ 0 & 0 & 2\sqrt{2} & 0 & 0 & 3 \\ 0 & 0 & 0 & 2\sqrt{2} & 3 & 0 \end{bmatrix}$$

Also, $\pm 5/2, \pm 3/2, \pm 1/2$ are 5-eigenvalues of A_{pq} with corresponding (unnor-

malized) eigenvectors given by:

$$\begin{bmatrix} 1 \\ 1 \\ \sqrt{5} \\ \sqrt{5} \\ \sqrt{10} \\ \sqrt{10} \end{bmatrix}, \begin{bmatrix} 1 \\ -1 \\ -\sqrt{5} \\ -\sqrt{5} \\ \sqrt{10} \\ -\sqrt{10} \end{bmatrix}, \begin{bmatrix} 1 \\ 1 \\ 3\sqrt{5} \\ -3\sqrt{5} \\ \sqrt{10}//5 \\ -\sqrt{10}/5 \end{bmatrix}, \begin{bmatrix} 1 \\ 1 \\ -3\sqrt{5} \\ -3\sqrt{5} \\ \sqrt{10}/5 \\ \sqrt{10}/5 \end{bmatrix}, \begin{bmatrix} 1 \\ 1 \\ \sqrt{5}/5 \\ \sqrt{5}/5 \\ -\sqrt{10}/5 \\ -\sqrt{10}/5 \end{bmatrix}, \begin{bmatrix} 1 \\ -1 \\ -\sqrt{5}/5 \\ \sqrt{5}/5 \\ -\sqrt{10}/5 \\ \sqrt{10}/5 \end{bmatrix} \quad \square$$

Continuing the pattern of these examples we have the following:

Conjecture. *Let S_n^{pq} be the n-subspace for A_{pq} generated by*

$$\left\{ \left| p^i q^j \right\rangle : i + j = n \right\}$$

Then $\pm n/2, \pm(n-2)/2, \pm(n-4)/2 \ldots$ are n-eigenvalues of A_{pq} with corresponding eigenvectors in S_n^{pq}.

The proof of this conjecture could be quite complicated. As a first step, we consider the case when n is even and the odd case is similar. In the given basis, A_{pq} is represented by the following self-adjoint matrix

$$\begin{bmatrix}
0 & 0 & \sqrt{n} & 0 & 0 & 0 & 0 & \cdots & 0 & 0 & 0 \\
0 & 0 & 0 & \sqrt{n} & 0 & 0 & 0 & \cdots & 0 & 0 & 0 \\
\sqrt{n} & 0 & 0 & 0 & \sqrt{2(n-1)} & 0 & 0 & \cdots & 0 & 0 & 0 \\
0 & \sqrt{n} & 0 & 0 & 0 & \sqrt{2(n-1)} & 0 & \cdots & . & . & . \\
0 & 0 & \sqrt{2(n-1)} & 0 & 0 & 0 & \sqrt{3(n-2)} & \cdots & . & . & . \\
0 & 0 & 0 & \sqrt{2(n-1)} & 0 & 0 & 0 & \cdots & . & . & . \\
0 & 0 & 0 & 0 & \sqrt{3(n-2)} & 0 & 0 & & & & \\
0 & 0 & 0 & 0 & 0 & \sqrt{3(n-2)} & 0 & & & & \\
0 & 0 & 0 & 0 & 0 & 0 & \sqrt{4(n-3)} & & & & \\
\vdots & \vdots & \vdots & \vdots & \vdots & \vdots & \vdots & & & & \\
0 & 0 & 0 & 0 & 0 & 0 & 0 & \cdots & 0 & \frac{1}{2}\sqrt{(n-2)(n+4)} & 0 \\
0 & 0 & 0 & 0 & 0 & 0 & 0 & \cdots & 0 & 0 & \frac{1}{2}\sqrt{n(n+2)} \\
0 & 0 & 0 & 0 & 0 & 0 & 0 & \cdots & 0 & 0 & \frac{1}{2}\sqrt{n(n+2)} \\
0 & 0 & 0 & 0 & 0 & 0 & 0 & \cdots & \frac{1}{2}\sqrt{n(n+2)} & \frac{1}{2}\sqrt{n(n+2)} & 0
\end{bmatrix}$$

We conjecture that the eigenvalues of this matrix are $\pm n/2, \pm(n-2)/2, \ldots, \pm 1, 0$. The corresponding eigenvectors become quite involved and we only display one of them. The reader can check that the (unnormalized) eigenvector corresponding to eigenvalue $n/2$ is

$$\left[\sqrt{\binom{n}{0}} \ \sqrt{\binom{n}{0}} \sqrt{\binom{n}{1}} \ \sqrt{\binom{n}{1}} \ \sqrt{\binom{n}{2}} \ \cdots \ \sqrt{\binom{n}{\frac{n}{2}-1}} \ \sqrt{\binom{n}{\frac{n}{2}-1}} \ \sqrt{\binom{n}{\frac{n}{2}}} \right]^T$$

where $\binom{n}{j} = \frac{n!}{(n-j)!j!}$.

Example 6. *We illustrate our conjecture for the case* $n = 12$. *The matrix for* A_{pq} *becomes:*

$$\begin{bmatrix}
0 & 0 & \sqrt{12} & 0 & 0 & 0 & 0 & 0 & 0 & 0 & 0 & 0 & 0 \\
0 & 0 & 0 & \sqrt{12} & 0 & 0 & 0 & 0 & 0 & 0 & 0 & 0 & 0 \\
\sqrt{12} & 0 & 0 & 0 & \sqrt{22} & 0 & 0 & 0 & 0 & 0 & 0 & 0 & 0 \\
0 & \sqrt{12} & 0 & 0 & 0 & \sqrt{22} & 0 & 0 & 0 & 0 & 0 & 0 & 0 \\
0 & 0 & \sqrt{22} & 0 & 0 & 0 & \sqrt{30} & 0 & 0 & 0 & 0 & 0 & 0 \\
0 & 0 & 0 & \sqrt{22} & 0 & 0 & 0 & \sqrt{30} & 0 & 0 & 0 & 0 & 0 \\
0 & 0 & 0 & 0 & \sqrt{30} & 0 & 0 & 0 & 6 & 0 & 0 & 0 & 0 \\
0 & 0 & 0 & 0 & 0 & \sqrt{30} & 0 & 0 & 0 & 6 & 0 & 0 & 0 \\
0 & 0 & 0 & 0 & 0 & 0 & 6 & 0 & 0 & 0 & \sqrt{40} & 0 & 0 \\
0 & 0 & 0 & 0 & 0 & 0 & 0 & 6 & 0 & 0 & 0 & \sqrt{40} & 0 \\
0 & 0 & 0 & 0 & 0 & 0 & 0 & 0 & \sqrt{40} & 0 & 0 & 0 & \sqrt{42} \\
0 & 0 & 0 & 0 & 0 & 0 & 0 & 0 & 0 & \sqrt{40} & 0 & 0 & \sqrt{42} \\
0 & 0 & 0 & 0 & 0 & 0 & 0 & 0 & 0 & 0 & \sqrt{42} & \sqrt{42} & 0
\end{bmatrix}$$

The 12*-eigenvalue* 6 *has corresponding (unnormalized) eigenvector:*

$$\begin{bmatrix} 1 & 1 & 2\sqrt{3} & 2\sqrt{3} & \sqrt{66} & \sqrt{66} & 2\sqrt{55} & 2\sqrt{55} & 3\sqrt{55} & 3\sqrt{55} & 6\sqrt{22} & 6\sqrt{22} & 2\sqrt{231} \end{bmatrix}^T \square$$

References

[1] Bisco A., D'Ariano G. and Perinotti P., *Special relativity in a discrete quantum universe*, arXiv:quant-ph 1503.01017v3 (2016).

[2] Crouse D., *On the nature of discrete space-time*, arXiv:quant-ph 1608.08506v1 (2016).

[3] Gudder S., *Discrete scalar quantum field theory*, arXiv:gen-ph 1610.07877v1 (2016).

[4] Gudder S., *Discrete spacetime quantum field theory*, arXiv:physics.gen-ph 1704.01639 (2017).

[5] Peskin M. and Schroeder D., *An Introduction to Quantum Field Theory*, Addison-Weseley, Reading, Mass. (1995).

[6] Veltman M., *Diagrammatica*, Cambridge University Press, Cambridge (1994).

In: Quantum Field Theory and Applications ISBN: 978-1-68507-930-7
Editor: N. Palerma

Chapter 4

PERTURVATIVE EXPANSION SERIES IN QED AND THE MUON G-2 ANOMALY – ONE OF THE OLDEST PROBLEMS IN QUANTUM FIELD THEORY AND OF THE LATEST PROBLEMS IN THE STANDARD MODEL*

Hidezumi Terazawa†
Center of Asia and Oceania for Science (CAOS), Tokyo, Japan
and
Midlands Academy of Business & Technology (MABT),
Leicester, UK

Abstract

Whether perturbative expansion series in quantum electrodynamics (QED) are convergent or not is discussed in detail, by taking the radiative corrections to the magnetic moment of a charged Dirac fermion such as the muon as an example. It is shown that they are

---

* Some of the contents of this Chapter have already been presented in the invited paper to the issue dedicated to the 29th anniversary of the NPCS Jounal, CAOS-Report-014(CAOS, Tokyo, 2018) September 1, 2018, published in Terazawa H., Nonlinear Phenomena in Complex Systems **21:3**, 268(2018).

†Corresponding Author's Email: terazawa@mrj.biglobe.ne.jp.

asymptotic expansion, convergent, or divergent series, depending on the fine-structure constant and the total number and masses of the charged particles. It is remarkable that their convergence not only constraints the fine-structure constant and the total number and masses of the fermions, but also forbids the existence of the magnetic monopole. It is also pointed out that the disagreement between the experimental value of the muon g-2 and the theoretical one predicted in the standard model, if it stayed, might be due either to the larger hadronic vacuum polarization contributions or to an unexpectedly large sum of the perturbative expansion series.

1. Introduction

In 1927, Dirac[1] invented quantum electrodynamics(QED), the relativistic quantum field theory describing electromagnetic interactions of electrons and photons. Then, it took almost two decades until Tomonaga[2] introduced "renormalization", the practical method to calculate physical quantities including radiative corrections (such as the anomalous magnetic moment of the electron[3]) by avoiding unphysical divergences. which often appear in higher-order terms in perturbative expansion series. However, although over seven decades have passed since then, we still do not know whether a sum of infinite series of perturbation expansions in QED is convergent or not, which seems to be one of the oldest problems in quantum field theory.
In 1951, Dyson[4] seemed to conjecture that perturbative expansion series in QED may be taken as asymptotic expansion series in general. We, however, strongly disagree with him as we can imagine that perturbative expansion series may be convergent or divergent, depending on the physical quantities and the values of expansion parameters such as the fine-structure constant and the ratios of the masses of charged particles. In this paper, we are going to discuss whether a sum of perturbative expansion series is convergent or not by taking the radiative corrections to the magnetic moment of a charged Dirac fermion such as the muon as an example.

Recently, one of the most intriguing problems in the standard model is that the theoretical value of the muon g-2 predicted in the standard model, $(g-2)_\mu^{SM} = (1165918.10 \pm 0.43) \times 10^{-9}$ (including the pure QED radiative corrections up to five-loops[5], the electroweak contributions up to two-loops[6], the hadronic vacuum polarization contributions[7], and the hadronic light-by-light scattering contributions[8]), disagrees with the experimental one[9], $(g-2)_\mu^{exp} = (1165920.61 \pm 0.41) \times 10^{-9}$, by over four $\sigma' s$:

$$(g-2)_\mu^{exp}/2 - (g-2)_\mu^{SM}/2 = (2.51 \pm 0.59) \times 10^{-9}$$

. Although this discrepancy, if it stayed [7], might indicate a valuable sign for new physics beyond the standard model, it might also pinpoint a possibility that a sum of the perturbative expansion series of higher orders would be unexpectedly large. This conjecture will be made at the end of this paper.

2. Expansion Parameters

The QED Lagrangian for N charged Dirac fermions with the charges and masses, e_i and m_i for $i = 1, 2, 3, ..., N$, is given by

$$L_{QED} = \sum_i \overline{\psi}_i [i\gamma^\mu(\partial_\mu - ie_i A_\mu) - m_i]\psi_i$$

$$-\frac{1}{4}(\partial_\mu A_\nu - \partial_\nu A_\mu)^2,$$

where ψ_i and A_μ are the charged fermion and photon fields.

It seems useful in QED to expand any S-matrix amplitudes into infinite series in terms of the fine-structure constant,

$$\alpha = e^2/4\pi,$$

where e is the electron charge magnitude, by assuming that the perturbation expansion series may converge. The smallness of the fine-structure constant, $\alpha = 1/137.035999139(31) \ll 1$[10], seems to justify the assumption. However, it is perfectly possible that the sum of the expansion series may diverge, depending on the physical quantities, the value of the fine-structure constant, and the total number and masses of the charged particles, which we will demonstrate in the next Section.

Also, we will introduce another expansion parameter which is very useful in discussing whether the perturative epansion series converge or not.

3. Muon Magnetic Moment

In 1962, Kinoshita[11] proved the theorem in QED that an unrenomlized amplitude does not diverge in the massless limit of a charged fermion in the loop. In 1968, I[12] introduced another expansion parameter in QED,

$$\beta = (2\alpha/3\pi)ln(m_\mu/m_e),$$

which is about 0.00825 experimentally[10], and calculated the anomalous magnetic moment of the muon of an arbitrary α^{n+1} order for $n = 1, 2, 3, ...$ as

$$(g-2)_\mu^{(n+1)}/2 = (\alpha/2\pi)\beta^n[1 + O(\alpha/\pi)],$$

in the leading order of β, by making use of the Kinoshita theorem.

Under the extreme condition of $1 \ll ln(m_\mu/m_e) \ll 3\pi/2\alpha$ or $\alpha/\pi \ll \beta \ll 1$, the leading series of radiative corrections to the muon g-2 can be summed up into the following form:

$$(g-2)_\mu/2 = (\alpha/2\pi)[1 + O(\alpha/\pi)]/(1-\beta).$$

Therefore, it seems tempting to expect that the pure QED radiative corrections up to arbitrary-number-loops can be summed up into a finite number. Also, in 1968, I[13] discussed all the hadronic vacuum polarization contributions to the anomalous magnetic moment of the muon and tried to obtain an upper bound on it. Furthermore, by assuming the finiteness of all the hadronic contributions, I even obtained constraints on the mass spectrum of hadrons.

Now we come to the main point in this Section:

1) Suppose the fine-structure constant should be so large as the β parameter might not be smaller than unity, the infinite sum of $\sum_n (g-2)_\mu^{(n+1)}/2$ for $n = 1, 2, 3, ...$ would diverge. In other words, the finiteness of the anomalous magnetic moment of the muon requires that the fine-structure constant, α, must be smaller than $3\pi/2ln(m_\mu/m_e)$, which is about 0.89 experimentally[10].

2) Suppose there should exist N charged Dirac fermions with the masses much smaller than the muon mass, the anomalous magnetic moment of the muon of α^{n+1} order would become approximately by N^n times larger than that of the same order but for $N = 1$. Therefore, the finiteness of the anomalous magnetic moment of the muon requires that the number of

charged Dirac fermions with the masses much smaller then the muon mass must be smaller than $1/\beta$, which is about 121.2 experimentally[10].
3) Suppose the electron mass should be much too much smaller or the muon mass should be much too much larger so that the ratio of m_μ/m_e should be astronomically large so that the β parameter might not be smaller than unity, the infinite sum of $\sum_n (g-2)_\mu^{(n+1)}/2$ for $n = 1, 2, 3, ...$ would diverge. In other words, the finiteness of the anomalous magnetic moment of the muon requires that the ratio of m_μ/m_e must be smaller than $exp(3\pi/2\alpha)$, which is about 10^{280} experimentally[10].

In concluding this Section, I would like to add that it may be worth trying to reform the perturbation expansion method from the beginning in QED or in quantum field theory in general as proposed by Shirkov[14] and by 't Hooft[15] over four decades ago.

4. Magnetic Monopole

In 1931, Dirac[16] introduced a quantum theory of the magnetic monopole and showed that the magnetic charge g must be quantized as

$$eg/4\pi = \pm n/2 (n = 1, 2, 3, ...).$$

In 1966, Schwinger[17] proposed a quantum theory of the dyon having both the electric and magnetic charges in which the quantization relation is different as

$$eg/4\pi = \pm n (n = 1, 2, 3, ...).$$

Later, Wu and Yang[18] introduced the magnetic monopole into the non-Abelian SU(2) gauge theory. Furthermore, in 1974, 't Hooft and Polyakov [19] independently showed that a non-singular monopole configuration can be constructed in the non-Abelian SU(2) gauge theory with adjoint scalar fields.

In either cases, the quantization condition indicates that the electromagnetic coupling constant of the monopole, if any, must be very large as $g^2/4\pi = 1/4\alpha$, which is about 34.3 experimentally[10]. Therefore, already in 1962, Cabibbo and Ferrari[20] pointed out that the magnetic monopole can be produced efficiently by photon-photon scatterings. However, it is not until 2006 or even more recently when Kurochkin *et al.* and others[21] proposed the production of magnetic monopoles by the two-photon process[22] at high energy pp collisions.

Now we come to the main point in this Section:
1) The cross section for the production of monopole-anti-monopole pairs by two photons to the lowest order of the monopole coupling constant is proportional to the $(g^2/4\pi)^2$, which is of the order of 10^3, while a radiative correction to the cross section of the first order is proportional to the $(g^2/4\pi)^4$, which is of the order of 10^6. It seems hard to imagine that the pertubation expansion series of radiative corrections to the cross section would converge. The convergence of the radiative corrections seems to forbid the existence of the magnetic monopole!
2) The cross section for the production of monopole-anti-monopole pairs with n soft-photons($n = 1, 2, 3, ...$) is proportional to $(g^2/4\pi)^{n+2}$, which is of the order of $10^{3(1+n/2)}$. It seems incredible to have an explosive emission of photons in the production of monopole-anti-monopole pairs by two photons once incident colliding photons reach the energy threshold for the pair-production of monopoles. This also seems to forbid the existence of the magnetic monopole
3) One way out of the above problems on the existence of the magnetic monopole is to make a hypothesis that the monopoles may exist but only in the form of "monopolium", which is a composite state of monopole-anti-monopole pairs tightly bounded by the strong Coulomb force between them. If this is the case, the monopolium would be produced by the two photon process as a resonance of two photons which decays instantly into thousands of photons. It may be called "photon-ball".

Conclusion

In the previous Sections, it has been argued that perturvative expansion series in QED is asymmptotic expansion, convergent, or divergent series, depending on the values of the fine-structure constant and the number and masses of charged particles. In other words, the convergence of the series gives some constrains on the values of the fine-structure constant and the total number and mass-ratios of charged particles.

In conclusion, it shoud be emphasized that the discrepancy between the present experimental and theoretical values of the muon anomalous magnetic moment[5-9], if it stayed[7], might be due either to the larger hadronic hadronic vacuum polarizaion contrbutions (as emphasized over a half century ago[7]) or to the unexpectedly large infinite sum of higher-order radiative corrections. In either case, the one of the latest problems in the standard model might be solved by the one of the oldest problems in quantum field theory, without leading to new physics beyond the standard model!

Acknowledgments

The author would like to thank Mrs.Nadya Columbus and Ms.Stella Rosa for inviting the author to participating in their publishing program for their hard-cover edited collection tentatively entitled: Quantum Field Theory and Applications, and to thank Professor Masaki Yasuè for correcting the original manuscript. He also wishes to thank Professor Toichiro Kinoshita for the useful helps and valuable collaborations during his stay as instructor-research associate in Laboratory of Nuclear Studies, Cornell University from 1969 to 1971 and for many helpful private communications which he has received since then, without which this work would never have been completed.

REFERENCES

[1] Dirac P. M. A., *Proc. Roy. Soc. Lond.* **A114**, 243(1927); **A117**, 610(1928); **A118**, 351(1928); **133**, 453(1932); Dirac P. A. M., Fock V. A., and Podolsky B., *Phys. Z. Sowj.* **2**, 216(1932); Bloch F., *Phys. Z. Sowj.* **5**, 301(1934).

[2] Tomonaga S., *Prog. Theor. Phys.* **1**, 27(1946); **2**, 101, 198(1947); **3**, 1, 101(1948).

[3] Schwinger J. S., *Phys. Rev.* **73**, 416(1948), predicting the first-order radiative correction to the magnetic moment of the electron of $(g-2)_e/2 = \alpha/2\pi$, which is called the Schwinger correction.

[4] Dyson F. J., *Phys. Rev.* **85**, 631(1951), and many private communications with Kinoshita T.

[5] Aoyama T., Hayakawa M., Kinoshita T., and Nio M., *Phys. Rev. Lett.* **109**, 111807(2012); Kurz A. *et al.*, *Phys. Rev. D* **92**, 073019(2015), predicting the pure QED radiative corrections up to five-loops as $(g-2)_\mu^{QED}/2 = 0.0011658471885(36)$. For the theoretical value of the electron g-2 in the standard model, see Aoyama T., Kinoshita T., and Nio M., Atoms **7(1)**, 28(2019), https://doi.org/10.3390/atoms010028/, predicting $(g-2)_e^{QED}/2 = 0.001159652181606(11)(12)(330)$ in which the experimental value of the fine-structure constant measured at Berkeley with Cs atoms [Parker R. H. *et al.*, *Science* **360**, 191(2018)] is used. This standard model prediction is 2.4σ higher than the experimental value [Hanneke D., Fogwell S., and Gabrielse G., *Phys. Rev. Lett.* **100**, 120801 (2008)]: $(g-2)_e^{exp}/2-(g-2)_e^{SM} = (-8.8 \pm 3.6)\times 10^{-13}$. However, the new standard model prediction, in which the new experimental value of the fine-structure constant measured at Laboratoire Kastler Brossel with Rb atoms of $\alpha^{-1} = 137.035999206(11)$ [L. Morel *et al.*, *Nature* **588**, 61(2020)] is used, is 2.6σ lower than the experimental value of $(4.8 \pm 3.0)\times 10^{-13}$. For the theoretical value of the muon g-2 in the standard model, see Aoyama T. *et al.*, *Phys. Rept.* **887**, 1(2020), predicting $(g-2)_\mu^{SM} = 0.00116591810(43)$.

[6] Knecht M., Peris S., Perrottet M., and de Rafael E., *JHEP* **0211**, 003(2002); Czarnecki A., Marciano W.J., and Vainstein A., *Phys. Rev. D* **67**, 073006(2003); **73**, 119901(E)(2006), predicting the electroweak contributions up to two-loops as $(g-2)_\mu^{EW}/2 = (154.0 \pm 1.0)\times 10^{-11}$. For the earliest proposals of electroweak-interaction models beyond the standard and the contributions to the muon g-2, see Kinoshita T., Pestieau J., Roy P., and Terazawa H., *Phys. Rev. D* **2**, 910(1970). Recently, there have appeared hundreds of references on the muon g-2 in various models beyond the standard in order to explain the muon g-2 anomaly. For some latest references, see Arcadi G., Calibbi L., Fedele M., and Mescia F., *Phys. Rev. Lett.* **127**, 061802(2021); Marzocca D. and Trifinopoulos S., *ibid.* **127**, 061803(2021), and references therein.

[7] Jegerlehner F., arXiv:1511.04473v2[hep-ph] 10 Dec 2015; Kurz A., Liu T., Marquard P., and Steinhauser M., *Phys. Lett. B* **734**, 144(2014), predicting the hadronic vacuum polarization contributions

as $(g-2)_\mu^{HVP}/2 = (6816.94 \pm 13.09) \times 10^{-11}$. See also Davier M., Hoecker A., Malaescu B., and Zhang Z., *Eur. Phys. J. C***80** 241(2020); Kashavarzi A., Nomura D., and Teubner T., *Phys. Rev. D* **101**, 014029(2020); Colangelo G., Hoferichter M., and Stoffer P., *JHEP* **02**, 006(2019); Hoferichter M., Hoid B.-L., and Kubis B., *JHEP* **08**, 137(2019); Aoyama T., *et al.*, *Phys. Rept.* **887**, 1(2020), predicting $(g-2)_\mu^{HVP}/2 = (6931 \pm 40) \times 10^{-11}$. Very recently, however, Borsanyi Sz. *et al.* have predicted $(g-2)_\mu^{HVP}/2 = 7075 \pm 55) \times 10^{-11}$ in lattice QCD, which favors the experimentally measured value over the results based on the dispersion relation[See Borsanyi Sz. *et al.*, *Nature* **593**, 51(2021)]. For the earliest attempts for estimating all the hadronic vacuum polarization contributions, leading to the upper bound of $(g-2)_\mu^{HVP}/2 \leq 2.2 \times 10^{-7}$, see Terazawa H., *Prog. Theor. Phys.* **39**, 1326(1968); *Phys. Rev.* **177**, 2159(1969), in which the possibility of having a large sum of the hadronic vacuum polarization contributions to explain the muon g-2 anomaly was emphasized more than five decades ago! For a related latest study of pion vector form factor and its contribution to the muon g-2, see Yi J.-Y., Wang Z.-Y., and Xiao C.W., arXiv:2107.09535v3 [hep-ph] 30 Jul 2021.

[8] Pauk V. and Vanderhaeghen M., *Eur. Phys.J. C***74**, 3008(2014); Jegerlehner F., arXiv:1511.04473v2[hep-ph] 10 Dec 2015; Knecht M., Narison S., Rabemananjara A., and Rabetiarivony D., arXiv:1808.03848v1[hep-ph] 11 Aug 2018; Hoferichter M., Hoid B.-L., Kubis B., Leupold S., and Schneider S.P., arXiv: 1808.04823v1[hep-ph] 14 Aug 2018, predicting the hadronic light-by-light scattering contributions as $(g-2)_\mu^{HLbL} = (91.7 \pm 8.1) \times 10^{-11}$. See also Danilkin I., Hoferichter M., and Stoffer P., arXiv:2105.01666v1[hep-ph]4 May 2021. For the earliest attempt for calculating the light-by-light scattering contributions, see Terazawa H., *Prog. Theor. Phys.* **38**, 863(1967).

[9] Bennett G.W. *et al.*[Muon g-2 Collaboration], *Phys. Rev. D* **73**, 072003(2006); Mohr P.J., Newell D.B., and Taylor B.N., *Rev. Mod. Phys.* **88**, 035009(2016), giving the experimental value of $(g-2)_\mu^{exp} = 0.001165920910(630)$, for the negative muon anomalous magnetic moment, and Abi B. *et al.* [Muon g-2 Collaboration], *Phys. Rev. Lett.* **126**, 141801(2021), giving $(g-2)_\mu^{exp} =$

$0.00116592040(54)$, for the positive muon anomalous moment. The new experimental average is $(g-2)_\mu^{exp} = 0.00116592061(41)$.

[10] Patrignani C. *et al.* (Particle Data Group), *Chin. Phys. C* **40**, 100001(2016).

[11] Kinoshita T., *J. Math. Phys.* **3**, 650(1962).

[12] Terazawa H., *Prog. Theor. Phys.* **40**, 830(1968).

[13] Terazawa H., *Prog. Theor. Phys.* **39**, 1326(1968); Phys. Rev. **177**, 2159(1969).

[14] Shirkov D.V., *Lett. Math. Phys.* **1**, 179(1976); *Lett. Nuovo Cim.* **18**, 452(1977). For his later works, see Shirkov D.V., in *Proc. International Conference "New Trends in High-Energy Physics*, Yalta (Crimea), 2006, edited by Bogolyubov P.N. *et al.* (Bogolyubov Institute for Theoretical Physics, Kiev, 2006), p.247, and references therein.

[15] 't Hooft G., in *The Whys of Subnuclear Physics* (the Proceedings of the 15th School, Erice, 1977), edited by Zichichi A. (Plenum, New York, 1979), p.943.

[16] Dirac P.A.M., *Proc. Roy. Soc. Lond.* **A133**, 60(1931); *Phys. Rev.* **74**, 817(1948).

[17] Schwinger J., *Phys. Rev.* **144**, 1087(1966); *Science* **165**, 757(1969).

[18] Wu T.T. andYang C.N., *Nucl. Phys.* **B107**, 365(1976).

[19] 't Hooft G., *Nucl. Phys. B* **79**, 276(1974); Polyakov A.M., *JETP Lett.* **20**, 194(1974).

[20] Cabibbo N. and Ferrari E., *Nuovo Cim.* **23**, 1147(1962).

[21] Kurochkin Yu., Satsunkevich I., Shoukavy Dz., Rusakovich N., and Kulchitsky Yu., *Mod. Phys. Lett.A* **21**, 2873(2006); Kurochkin Yu., Satsunkevich I., and Shoukavy Dz., in *Proc. 5th Bolyai-Gauss-Lobachevsky International Conference "Non-Euclidean Geometry in Modern Physics"*, Minsk, Belarus, 2006, edited by Kurochkin Yu.

and Red'kov V. (Institute of Physics, National Academy of Sciences of Belarus, Minsk, 2006), p.349; Satsunkevich I., Shoukavy Dz., and Tsiareshka P., in *Conf. Proc. C0908171*, 517(2009); Acharya B. *et al.*[MoEDAL Collaboration], *Int. J. Mod. Phys.* **A29**, 1430050(2014), arXiv:1405.7662v4[hep-ph] 28 Aug 2014; Baines S., Mavromatos N.E., Mitsou V.A., Pinfold J.L., and Santra A., arXiv:1808.08942v1 [hep-ph] 27 Aug 2018. For the latest experimental search for production of magnetic monopoles in Pb-Pb heavy-ion collisions at the LHC, see Acharya B. *et al.*, arXiv:2106.11933v1 [hep-ex] 22 Jun 2021.

[22] For a review, see, for example, Terazawa H., *Rev. Mod. Phys.* **45**, 615(1973).

Bibliography

A prelude to quantum field theory	
LCCN	2021034292
Type of material	Book
Personal name	Donoghue, John, 1950- author.
Main title	A prelude to quantum field theory / John Donoghue and Lorenzo Sorbo.
Published/Produced	Princeton: Princeton University Press, [2022]
ISBN	9780691223490 (hardback)
	9780691223483 (paperback)
	(ebook)
LC classification	QC174.45 .D66 2022
Related names	Sorbo, Lorenzo, 1973- author.
Summary	"A Prelude to Quantum Field Theory offers a short introduction to quantum field theory (QFT), a powerful framework for understanding particle behavior that is an essential tool across many subfields of physics. A subject that is typically taught at the graduate level in most physics departments, quantum field theory is a unification of standard quantum theories and special relativity, which depicts all particles as

	"excitations" that arise in underlying fields. It extends quantum mechanics, the modern theory of one or few particles, in a way that is useful for the analysis of many-particle systems in the real world. As it requires a different style of thinking from quantum mechanics, which is typically the undergraduate physics student's first encounter with the quantum world, many beginners struggle with the transition to quantum field theory, especially when working with traditional textbooks. Existing books on the subject often tend to be large, sophisticated, and complete; and an overwhelming wealth of information and technical detail makes it difficult for the novice to discern what is most important. This book is a concise, friendly entrée for QFT-beginners, guiding the reader from the style of quantum mechanical thinking to that of QFT, and distilling the key ideas without a welter of unnecessary detail. In contrast with standard texts, which are predominantly particle physics-centric, this book is designed to be "subfield-neutral" - usable by students of any background and interest, and easily adaptable in a course setting according to instructors' preferences. The authors' conviction is that QFT is a core element of physics that should be understood by all PhD physicists-but that developing an appreciation for it does not require digesting a large, encyclopedic volume"-- Provided by publisher.
	"A concise, beginner-friendly introduction to quantum field theory. Quantum field theory is a powerful framework that extends quantum mechanics in ways that are essential in many

	modern applications. While it is the fundamental formalism for the study of many areas of physics, quantum field theory requires a different way of thinking, and many newcomers to the subject struggle with the transition from quantum mechanics. A Prelude to Quantum Field Theory introduces the key concepts of quantum field theory in a brief and accessible manner while never sacrificing mathematical rigor. The result is an easy-to-use textbook that distills the most general properties of the theory without overwhelming beginning students with more advanced applications. Bridges quantum mechanics and quantum field theory, emphasizing analogies and differences. Emphasizes a "quantum field theoretical mindset" while maintaining mathematical rigor. Obtains quantum fields as the continuum limit of a quantized system of many particles. Highlights the correspondence between wave function-fundamental in quantum mechanics-and the formalism of second quantization used in quantum field theory. Provides a step-by-step derivation of Feynman rules for the perturbative study of interacting theories. Introduces students to renormalization, path integrals techniques, and more. Discusses more modern topics like effective field theories. Ideal for both undergraduate and graduate students. Proven in the classroom"-- Provided by publisher.
Subjects	Quantum field theory.
	Science / Physics / Quantum Theory
Notes	Includes bibliographical references and index.

Additional formats	Online version: Donoghue, John, 1950- Prelude to quantum field theory Princeton: Princeton University Press, [2022] 9780691223506 (DLC) 2021034293

A prelude to quantum field theory	
LCCN	2021034293
Type of material	Book
Personal name	Donoghue, John, 1950- author.
Main title	A prelude to quantum field theory / John Donoghue and Lorenzo Sorbo.
Published/Produced	Princeton: Princeton University Press, [2022]
Description	1 online resource
ISBN	9780691223506 (ebook)
	(hardback)
	(paperback)
LC classification	QC174.45
Related names	Sorbo, Lorenzo, 1973- author.
Summary	"A Prelude to Quantum Field Theory offers a short introduction to quantum field theory (QFT), a powerful framework for understanding particle behavior that is an essential tool across many subfields of physics. A subject that is typically taught at the graduate level in most physics departments, quantum field theory is a unification of standard quantum theories and special relativity, which depicts all particles as "excitations" that arise in underlying fields. It extends quantum mechanics, the modern theory of one or few particles, in a way that is useful for the analysis of many-particle systems in the real world. As it requires a different style of thinking from quantum mechanics, which is typically the

	undergraduate physics student's first encounter with the quantum world, many beginners struggle with the transition to quantum field theory, especially when working with traditional textbooks. Existing books on the subject often tend to be large, sophisticated, and complete; and an overwhelming wealth of information and technical detail makes it difficult for the novice to discern what is most important. This book is a concise, friendly entrée for QFT-beginners, guiding the reader from the style of quantum mechanical thinking to that of QFT, and distilling the key ideas without a welter of unnecessary detail. In contrast with standard texts, which are predominantly particle physics-centric, this book is designed to be "subfield-neutral" - usable by students of any background and interest, and easily adaptable in a course setting according to instructors' preferences. The authors' conviction is that QFT is a core element of physics that should be understood by all PhD physicists-but that developing an appreciation for it does not require digesting a large, encyclopedic volume"-- Provided by publisher.
	"A concise, beginner-friendly introduction to quantum field theory. Quantum field theory is a powerful framework that extends quantum mechanics in ways that are essential in many modern applications. While it is the fundamental formalism for the study of many areas of physics, quantum field theory requires a different way of thinking, and many newcomers to the subject struggle with the transition from quantum mechanics. A Prelude to Quantum Field Theory

	introduces the key concepts of quantum field theory in a brief and accessible manner while never sacrificing mathematical rigor. The result is an easy-to-use textbook that distills the most general properties of the theory without overwhelming beginning students with more advanced applications. Bridges quantum mechanics and quantum field theory, emphasizing analogies and differences. Emphasizes a "quantum field theoretical mindset" while maintaining mathematical rigor. Obtains quantum fields as the continuum limit of a quantized system of many particles. Highlights the correspondence between wave function-fundamental in quantum mechanics-and the formalism of second quantization used in quantum field theory. Provides a step-by-step derivation of Feynman rules for the perturbative study of interacting theories. Introduces students to renormalization, path integrals techniques, and more. Discusses more modern topics like effective field theories. Ideal for both undergraduate and graduate students. Proven in the classroom"-- Provided by publisher.
Subjects	Quantum field theory.
	Science / Physics / Quantum Theory
Notes	Includes bibliographical references and index.
	Description based on print version record and CIP data provided by publisher; resource not viewed.
Additional formats	Print version: Donoghue, John, 1950- Prelude to quantum field theory Princeton: Princeton University Press, [2022] 9780691223490 (DLC) 2021034292

A short introduction to string theory	
LCCN	2021062166
Type of material	Book
Personal name	Mohaupt, Thomas, author.
Main title	A short introduction to string theory / Thomas Mohaupt, University of Liverpool.
Published/Produced	Cambridge, United Kingdom; New York, NY: Cambridge University Press, 2022.
ISBN	9781108481380 (hardback)
	(epub)
LC classification	QC794.6.S85 M64 2022
Summary	"Suitable for graduate students in physics and mathematics, this book presents a concise and pedagogical introduction to string theory. It focuses on explaining the key concepts of string theory, such as bosonic strings, D-branes, supersymmetry, and superstrings and on clarifying the relationship between particles, fields, and strings without assuming an advanced background in particle theory or quantum field theory, thus making it widely accessible to interested readers from a range of backgrounds. Important ideas underpinning current research, such as partition functions, compactification, gauge symmetries, and T-duality are analysed both from the world-sheet (conformal field theory) and the space-time (effective field theory) perspectives. Ideal for either self-study or a one semester graduate course, A Short Introduction to String Theory is an essential resource for students studying string theory, containing examples and homework problems to develop understanding, with fully worked

	solutions available to instructors"-- Provided by publisher.
Subjects	String models.
	Science / Physics / Mathematical & Computational
Notes	Includes bibliographical references and index.
Additional formats	Online version: Mohaupt, Thomas. Short introduction to string theory Cambridge, United Kingdom; New York, NY: Cambridge University Press, 2022 9781108611619 (DLC) 2021062167

Advanced topics in quantum field theory: a lecture course	
LCCN	2021029678
Type of material	Book
Personal name	Shifman, Mikhail A., author.
Main title	Advanced topics in quantum field theory: a lecture course / Mikhail Shifman, University of Minnesota.
Edition	Second edition.
Published/Produced	Cambridge, UK; New York, NY: Cambridge University Press, 2022.
ISBN	9781108840422 (hardback)
	(epub)
LC classification	QC174.46 .S55 2021
Summary	"Quantum field theory is the basis of our modern description of physical phenomena at the fundamental level. This systematic and comprehensive text emphasizes nonperturbative phenomena and supersymmetry. It includes a thorough discussion of various phases of gauge theories, extended objects and their quantization, and global supersymmetry from a modern perspective. This Second Edition is revised to

	include topics developed in the last decade, including higher-form global symmetries and their applications, anomalies in supersymmetric theories beyond Ferrara-Zumino, and non-Abelian supersymmetric vortex strings. A new final part is added, presenting more than 90 problems with detailed solutions, allowing students to check their understanding of the acquired knowledge and providing extra details to supplement the main text descriptions. This an indispensable book for graduate students and researchers in theoretical physics"-- Provided by publisher.
Subjects	Quantum field theory.
Notes	Includes bibliographical references and index.
Additional formats	Online version: Shifman, Mikhail. Advanced topics in quantum field theory 2. New York: Cambridge University Press, 2021 9781108885911 (DLC) 2021029679

Advanced topics in quantum field theory: a lecture course	
LCCN	2021029679
Type of material	Book
Personal name	Shifman, Mikhail A., author.
Main title	Advanced topics in quantum field theory: a lecture course / Mikhail Shifman, University of Minnesota.
Edition	Second edition.
Published/Produced	Cambridge, UK; New York, NY: Cambridge University Press, 2022.
Description	1 online resource
ISBN	9781108885911 (epub)
	(hardback)

LC classification	QC174.46
Summary	"Quantum field theory is the basis of our modern description of physical phenomena at the fundamental level. This systematic and comprehensive text emphasizes nonperturbative phenomena and supersymmetry. It includes a thorough discussion of various phases of gauge theories, extended objects and their quantization, and global supersymmetry from a modern perspective. This Second Edition is revised to include topics developed in the last decade, including higher-form global symmetries and their applications, anomalies in supersymmetric theories beyond Ferrara-Zumino, and non-Abelian supersymmetric vortex strings. A new final part is added, presenting more than 90 problems with detailed solutions, allowing students to check their understanding of the acquired knowledge and providing extra details to supplement the main text descriptions. This an indispensable book for graduate students and researchers in theoretical physics"-- Provided by publisher.
Subjects	Quantum field theory.
Notes	Includes bibliographical references and index.
	Description based on print version record and CIP data provided by publisher; resource not viewed.
Additional formats	Print version: Shifman, Mikhail A. Advanced topics in quantum field theory Second edition. Cambridge, UK; New York, NY: Cambridge University Press, 2022 9781108840422 (DLC) 2021029678

Amplitudes, Hodge theory and ramification: from periods and motives to Feynman amplitudes: 2014 Clay Mathematics Institute Summer School: periods and motives: Feynman amplitudes in the 21st century, June 30-July 25, 2014, Instituto de Ciencias Matemáticas, Madrid, Spain	
LCCN	2020001922
Type of material	Book
Corporate name	Clay Mathematics Institute. Summer School (2014: Madrid, Spain), author.
Main title	Amplitudes, Hodge theory and ramification: from periods and motives to Feynman amplitudes: 2014 Clay Mathematics Institute Summer School: periods and motives: Feynman amplitudes in the 21st century, June 30-July 25, 2014, Instituto de Ciencias Matemáticas, Madrid, Spain / K. Ebrahimi-Fard, J.I. Burgos Gil, D. Manchon, editors.
Published/Produced	Providence, RI: Published by the American Mathematical Society for the Clay Mathematics Institute, [2020]
Description	xiv, 229 pages: illustrations; 26 cm
ISBN	9781470443290 (softcover)
	(ebook)
LC classification	QC19.2 .C586 2020
Related names	Ebrahimi-Fard, Kurusch, 1973- editor.
	Burgos Gil, José I. (José Ignacio), 1962- editor.
	Manchon, Dominique, 1962- editor.
Contents	Foreword / Yuri I. Manin -- Feynman integrals in mathematics and physics / Spencer Bloch -- Feynman integrals and periods in configuration spaces / Özgür Ceyhan and Matilde Marcolli -- Introductory course on ℓ-adic sheaves and their

	ramification theory on curves / Lars Kindler and Kay Rulling
Subjects	Mathematical physics--Congresses.
	Quantum field theory--Congresses.
	Feynman integrals--Congresses.
	Hodge theory--Congresses.
	Motives (Mathematics)--Congresses.
	Geometry, Algebraic--Congresses.
	Algebraic number theory--Congresses.
	Quantum theory -- General mathematical topics and methods in quantum theory -- Feynman integrals and graphs; applications of algebraic topology and algebraic geometry.
	Algebraic geometry -- Cycles and subschemes -- (Equivariant) Chow groups and rings; motives.
	Number theory -- Algebraic number theory: local and \$p\$-adic fields -- Ramification and extension theory.
Notes	Includes bibliographical references.
Series	Clay mathematics proceedings, 1534-6455; volume 21

Applications of field theory methods in statistical physics of nonequilibrium systems	
LCCN	2021005227
Type of material	Book
Personal name	Lev, Bohdan, author.
Main title	Applications of field theory methods in statistical physics of nonequilibrium systems / Bohdan Lev, National Academy of Science of Ukraine, Ukraine, Anatoly Zagorodny, National Academy of Science of Ukraine, Ukraine.
Published/Produced	New Jersey: World Scientific, [2021]

Description	1 online resource
ISBN	9789811229985 (ebook)
	(hardcover)
LC classification	QC174.7
Related names	Zagorodny, A., author.
Summary	"This book formulates a unified approach to the description of many-particle systems combining the methods of statistical physics and quantum field theory. The benefits of such an approach are in the description of phase transitions during the formation of new spatially inhomogeneous phases, as well in describing quasi-equilibrium systems with spatially inhomogeneous particle distributions (for example, self-gravitating systems) and metastable states. The validity of the methods used in the statistical description of many-particle systems and models (theory of phase transitions included) is discussed and compared. The idea of using the quantum field theory approach and related topics (path integration, saddle-point and stationary-phase methods, Hubbard-Stratonovich transformation, mean-field theory, and functional integrals) is described in detail to facilitate further understanding and explore more applications. To some extent, the book could be treated as a brief encyclopedia of methods applicable to the statistical description of spatially inhomogeneous equilibrium and metastable particle distributions. Additionally, the general approach is not only formulated, but also applied to solve various practically important problems (gravitating gas, Coulomb-like systems, dusty plasmas, thermodynamics of cellular structures, non-

	uniform dynamics of gravitating systems, etc.)"-- Provided by publisher.
Subjects	Statistical physics.
	Field theory (Physics)
Notes	Includes bibliographical references and index.
	Description based on print version record and CIP data provided by publisher; resource not viewed.
Additional formats	Print version: Lev, Bohdan. Applications of field theory methods in statistical physics of nonequilibrium systems New Jersey: World Scientific, [2021] 9789811229978 (DLC) 2021005226

Applications of field theory methods in statistical physics of nonequilibrium systems	
LCCN	2021005226
Type of material	Book
Personal name	Lev, Bohdan, author.
Main title	Applications of field theory methods in statistical physics of nonequilibrium systems / Bohdan Lev, National Academy of Science of Ukraine, Ukraine, Anatoly Zagorodny, National Academy of Science of Ukraine, Ukraine.
Published/Produced	New Jersey: World Scientific, [2021]
ISBN	9789811229978 (hardcover)
	(ebook)
LC classification	QC174.7 .L47 2021
Related names	Zagorodny, A., author.
Summary	"This book formulates a unified approach to the description of many-particle systems combining the methods of statistical physics and quantum field theory. The benefits of such an approach are

	in the description of phase transitions during the formation of new spatially inhomogeneous phases, as well in describing quasi-equilibrium systems with spatially inhomogeneous particle distributions (for example, self-gravitating systems) and metastable states. The validity of the methods used in the statistical description of many-particle systems and models (theory of phase transitions included) is discussed and compared. The idea of using the quantum field theory approach and related topics (path integration, saddle-point and stationary-phase methods, Hubbard-Stratonovich transformation, mean-field theory, and functional integrals) is described in detail to facilitate further understanding and explore more applications. To some extent, the book could be treated as a brief encyclopedia of methods applicable to the statistical description of spatially inhomogeneous equilibrium and metastable particle distributions. Additionally, the general approach is not only formulated, but also applied to solve various practically important problems (gravitating gas, Coulomb-like systems, dusty plasmas, thermodynamics of cellular structures, non-uniform dynamics of gravitating systems, etc.)"--Provided by publisher.
Subjects	Statistical physics.
	Field theory (Physics)
Notes	Includes bibliographical references and index.
Additional formats	Online version: Lev, Bohdan, Applications of field theory methods in statistical physics of nonequilibrium systems New Jersey: World

	Scientific, 2021. 9789811229985 (DLC) 2021005227

Asymptotic behavior: an overview	
LCCN	2019957631
Type of material	Book
Main title	Asymptotic behavior: an overview / Steve P. Riley, editor.
Published/Produced	New York: Nova Science Publishers, [2020]
Description	ix, 135 pages; 23 cm.
ISBN	9781536172225 (paperback)
	(adobe pdf)
LC classification	QA372 .A89 2020
Related names	Riley, Steve P., editor.
Summary	"Asymptotic Behavior: An Overview is designed to provide the reader with an exposition of some aspects of the oscillation theory of first order delay partial dynamic equations on time scales. Oscillation theory of differential equations, originated from the monumental paper of C. Sturm published in 1836, has now been recognized as an important branch of mathematical analysis from both theoretical and practical viewpoints. Asymptotic behavior in the deep Euclidean region of momenta for four-dimensional models of quantum field theory is studied through the system of Schwinger-Dyson equations. This system is truncated by a sequence of n-particle approximations in which n goes into the complete system of Schwinger-Dyson equations. Lastly, the authors discuss the exact analytical solution of the Schrödinger equation corresponding to the hydrogen atom

	confined by four spherical potentials: infinite potential, parabolic potential, constant potential, and dielectric continuum"-- Provided by publisher.
Subjects	Differential equations.
Notes	Includes bibliographical references and index.
Series	Mathematics research developments

Chiral differential operators via quantization of the holomorphic s-model	
LCCN	2021395434
Type of material	Book
Personal name	Gorbounov, Vassily, author.
Main title	Chiral differential operators via quantization of the holomorphic s-model / Vassily Gorbounov, Owen Gwilliam & Brian Williams.
Published/Produced	Paris, France: Société mathématique de France, 2020.
Description	ix, 210 pages; 24 cm
Rights advisory	Current Copyright Fee: GBP20.00 0.
ISBN	9782856299203
	2856299202
Related names	Gwilliam, Owen, author.
	Williams, Brian (Brian R.), author.
Subjects	Quantum field theory.
	Vertex operator algebras.
	Differential operators.
	Differential operators.
	Quantum field theory.
	Vertex operator algebras.
Notes	Includes bibliographical references (pages 207-210).

	Text in English with summaries in English and French.
Series	Astérisque, 0303-1179; numéro 419, 2020
	Astérisque; 419 0303-1179

Combinatorial physics: combinatorics, quantum field theory, and quantum gravity models	
LCCN	2021932400
Type of material	Book
Personal name	Tanasa, Adrian, author.
Main title	Combinatorial physics: combinatorics, quantum field theory, and quantum gravity models / Adrian Tanasa, University of Bordeaux, France.
Published/Produced	Oxford: Oxford University Press, 2021.
	©2021
Description	xi, 396 pages: illustrations; 25 cm
ISBN	9780192895493
	0192895494
LC classification	QC174.45 T355 2021
Subjects	Combinatorial analysis.
	Quantum field theory.
	Combinatorial analysis.
	Quantum theory.
Notes	Includes bibliographical references (p. 383-394) and index.

Completion and unification of quantum mechanics with Einstein's GR ideas Part III: Advances, revisions and conclusions	
LCCN	2020519397
Type of material	Book
Personal name	Majkić, Zoran author.
Main title	Completion and unification of quantum mechanics with Einstein's GR ideas Part III:

	Advances, revisions and conclusions / Zoran Majkíc.
Published/Produced	New York: Nova Science Publishers, [2020]
	©2020
Description	xlii, 413 pages: illustrations; 25 cm.
ISBN	9781536172003
Subjects	Quantum field theory.
	General relativity (Physics)
Notes	Includes bibliographical references and index.
Series	Classical and quantum mechanics

Diagrammatics: lectures on selected problems in condensed matter theory	
LCCN	2019039110
Type of material	Book
Personal name	Sadovskiĭ, M. V. (Mikhail Vissarionovich), 1948- author.
Main title	Diagrammatics: lectures on selected problems in condensed matter theory / Michael V. Sadovskii, Russian Academy of Sciences, Russia.
Edition	2nd edition.
Published/Produced	New Jersey: World Scientific, [2020]
ISBN	9789811212208 (hardcover)
	(ebook)
LC classification	QC173.454 .S23 2020
Summary	"The introduction of quantum field theory methods has led to a kind of "revolution" in condensed matter theory, resulting in the increased importance of Feynman diagrams or diagram technique. So, it has now become imperative for professionals in condensed matter theory to have a thorough knowledge of this method. The book is intended to teach students,

	postdocs and young theorists to use diagrammatic quantum field theory methods applied to different problems of modern condensed matter theory, using specific examples of such problems. This latest edition is extended by the inclusion of some new material on superconductivity and diagram combinatorics"-- Provided by publisher.
Subjects	Condensed matter.
	Quantum field theory.
	Feynman diagrams.
Notes	Includes bibliographical references and index.

Elementary particle physics: the standard theory	
LCCN	2021936417
Type of material	Book
Personal name	Iliopoulos, John, author.
Main title	Elementary particle physics: the standard theory / John Iliopoulos, Theodore N. Tomaras.
Published/Produced	New York: Oxford University Press, 2021.
ISBN	9780192844200 (hardback)
	9780192844217 (paperback)
	(ebook)
Related names	Tomaras, Theodore N., author.
Summary	"Determining the nature of matter's smallest constituents as well as the interactions among them is the subject of a branch of fundamental physics called "The Physics of Elementary Particles". It is the subject of this book. During the last decades this field has gone through a phase transition. It culminated in the formulation of a new theoretical scheme, known as "The Standard Model", which brought profound changes in our ways of thinking and

	understanding nature's fundamental forces. Its agreement with experiment is impressive, to the extent that we should no more talk about "The Standard Model" but instead "The Standard Theory". This new vision is based on geometry, the interactions are required to satisfy a certain geometrical principle. In the physicists' jargon this principle is called "gauge invariance", in mathematics it is a concept of differential geometry. It is the purpose of this book to present and explain this modern viewpoint to a readership of well-motivated undergraduate students. We propose to guide the reader to the more advanced concepts of Gauge Symmetry, Quantum Field Theory and the phenomenon of spontaneous symmetry breaking through concrete physical examples. The presentation of the techniques required for Particle Physics is self-contained and the mathematics is kept at the absolutely necessary level. The reader is invited to join the glorious parade of the theoretical advances and experimental discoveries of the last decades which established our current view. Our ambition is to make this fascinating subject accessible to undergraduate students and, hopefully, to motivate them to study it further"-- Provided by publisher.

Forks in the road: a life in physics	
LCCN	2021028052
Type of material	Book
Personal name	Deser, S. (Stanley), author.
Main title	Forks in the road: a life in physics / Stanley Deser.

Published/Produced	New Jersey: World Scientific, [2022]
Description	viii, 155 pages, 11 pages of plates: illustrations (chiefly color); 24 cm
ISBN	9789811234187 (hardcover)
	9789811235665 (paperback)
	(ebook for institutions)
	(ebook for individuals)
LC classification	QC16.D48 A3 2022
Summary	"A scientific autobiography by Stanley Deser. Stanley Deser is a preeminent theoretical physicist who made monumental contributions to general relativity, quantum field theory and high energy physics; he is a co-creator of supergravity. This is his personal story, intended for a broad, scientifically curious audience, with emphasis on the historic figures that defined the modern aspects of the field. Beginning with an account of his early life in Europe during the fateful period leading up to WW2, it continues with his family's dramatic escape from the Nazis through their arrival to the US. His education at public institutions including Brooklyn College nurtured his love of physics from an early age. He earned his PhD at Harvard and spent fruitful postdoc years at the Institute for Advanced Study and the Niels Bohr Institute, where he met many of the luminaries of the field. Then followed a long career at Brandeis University and many visits to foreign institutions. His work earned him many awards and led to exotic experiences detailed in the later chapters. The appendices contain semi-technical descriptions of some essential physics, as well as a more general commentary about the role of physics and

	physicists in understanding the universe"--Provided by publisher.
Contents	Early days: rootless cosmopolitan -- America -- College days -- Graduate school -- The institute for advanced study -- The Bohr Institute -- America again -- Boston - adulthood -- Academia: (UN)steady state -- Innocents abroad -- A big year -- Oxford et al. -- The eighties -- The iron curtain I -- The iron curtain II -- Tiananmen -- A last sabbatical -- Meetings and memorials -- Distinctions -- Festspiel.
Subjects	Deser, S. (Stanley)
	Physicists--United States--Biography.
Notes	Includes bibliographical references and index.
Additional formats	Online version: Deser, S. (Stanley) Forks in the road New Jersey: World Scientific, [2021] 9789811234194 (DLC) 2021028053

Forks in the road: a life in physics	
LCCN	2021028053
Type of material	Book
Personal name	Deser, S. (Stanley), author.
Main title	Forks in the road: a life in physics / Stanley Deser.
Published/Produced	New Jersey: World Scientific, [2021]
Description	1 online resource
ISBN	9789811234200 (ebook for individuals)
	(hardcover)
	9789811234194 (ebook for institutions)
	(paperback)
LC classification	QC16.D48
Summary	"A scientific autobiography by Stanley Deser. Stanley Deser is a preeminent theoretical

	physicist who made monumental contributions to general relativity, quantum field theory and high energy physics; he is a co-creator of supergravity. This is his personal story, intended for a broad, scientifically curious audience, with emphasis on the historic figures that defined the modern aspects of the field. Beginning with an account of his early life in Europe during the fateful period leading up to WW2, it continues with his family's dramatic escape from the Nazis through their arrival to the US. His education at public institutions including Brooklyn College nurtured his love of physics from an early age. He earned his PhD at Harvard and spent fruitful postdoc years at the Institute for Advanced Study and the Niels Bohr Institute, where he met many of the luminaries of the field. Then followed a long career at Brandeis University and many visits to foreign institutions. His work earned him many awards and led to exotic experiences detailed in the later chapters. The appendices contain semi-technical descriptions of some essential physics, as well as a more general commentary about the role of physics and physicists in understanding the universe"--Provided by publisher.
Contents	Early days: rootless cosmopolitan -- America -- College days -- Graduate school -- The institute for advanced study -- The Bohr Institute -- America again -- Boston - adulthood -- Academia: (UN)steady state -- Innocents abroad -- A big year -- Oxford et al. -- The eighties -- The iron curtain I -- The iron curtain II --

	Tiananmen -- A last sabbatical -- Meetings and memorials -- Distinctions -- Festspiel.
Subjects	Deser, S. (Stanley)
	Physicists--United States--Biography.
Notes	Includes bibliographical references and index.
	Description based on print version record and CIP data provided by publisher; resource not viewed.
Additional formats	Print version: Deser, S. Forks in the road New Jersey: World Scientific, [2021] 9789811234187 (DLC) 2021028052

Foundations of modern physics	
LCCN	2020055431
Type of material	Book
Personal name	Weinberg, Steven, 1933- author.
Main title	Foundations of modern physics / Steven Weinberg, The University of Texas at Austin.
Published/Produced	New York: Cambridge University Press, 2021.
ISBN	9781108841764 (hardback)
	(epub)
LC classification	QC21.3 .W345 2021
Summary	"In addition to his ground-breaking research, Nobel Laureate Steven Weinberg is known for a series of highly praised texts on various aspects of physics, combining exceptional physical insight with a gift for clear exposition. Describing the foundations of modern physics in their historical context and with some new derivations, Weinberg introduces topics ranging from early applications of atomic theory through thermodynamics, statistical mechanics, transport theory, special relativity, quantum mechanics,

	nuclear physics, and quantum field theory. This volume provides the basis for advanced undergraduate and graduate physics courses as well as being a handy introduction to aspects of modern physics for working scientists"-- Provided by publisher.
Subjects	Physics.
Notes	Includes bibliographical references and indexes.
Additional formats	Online version: Weinberg, Steven, Foundations of modern physics 1. New York: Cambridge University Press, 2021. 9781108894845 (DLC) 2020055432

Foundations of modern physics	
LCCN	2020055432
Type of material	Book
Personal name	Weinberg, Steven, 1933- author.
Main title	Foundations of modern physics / Steven Weinberg, The University of Texas at Austin.
Published/Produced	New York: Cambridge University Press, 2021.
Description	1 online resource
ISBN	9781108894845 (epub)
	(hardback)
LC classification	QC21.3
Summary	"In addition to his ground-breaking research, Nobel Laureate Steven Weinberg is known for a series of highly praised texts on various aspects of physics, combining exceptional physical insight with a a gift for clear exposition. Describing the foundations of modern physics in their historical context and with some new derivations, Weinberg introduces topics ranging from early applications of atomic theory through

	thermodynamics, statistical mechanics, transport theory, special relativity, quantum mechanics, nuclear physics, and quantum field theory. This volume provides the basis for advanced undergraduate and graduate physics courses as well as being a handy introduction to aspects of modern physics for working scientists"--Provided by publisher.
Subjects	Physics.
Notes	Includes bibliographical references and indexes.
	Description based on print version record and CIP data provided by publisher; resource not viewed.
Additional formats	Print version: Weinberg, Steven, 1933- Foundations of modern physics New York: Cambridge University Press, 2021. 9781108841764 (DLC) 2020055431

Foundations of quantum field theory	
LCCN	2020037035
Type of material	Book
Personal name	Rothe, Klaus D. (Klaus Dieter) author.
Main title	Foundations of quantum field theory / Klaus D. Rothem, University of Heidelberg, Germany.
Published/Produced	Hackensack: World Scientific, [2021]
ISBN	9789811221927 (hardcover)
	9789811223006 (paperback)
	(ebook)
LC classification	QC174.24.R4 R68 2020
Summary	"Based on a two-semester course held at the University of Heidelberg, Germany, this book provides an adequate resource for the lecturer and the student. The contents are primarily aimed at

	graduate students who wish to learn about the fundamental concepts behind constructing a Relativistic Quantum Theory of particles and fields. So it provides a comprehensive foundation for the extension to Quantum Chromodynamics and Weak Interactions, that are not included in this book"-- Provided by publisher.
Subjects	Relativistic quantum theory.
Notes	Includes bibliographical references and index.
Series	World scientific lecture notes in physics, 1793-1436; vol. 84

From the past to the future: the legacy of Lev Lipatov	
LCCN	2021049422
Type of material	Book
Main title	From the past to the future: the legacy of Lev Lipatov / editors, Joachim Bartels, Victor Fadin, Eugene Levin, Aharon Levy, Victor Kim, Agustín Sabio-Vera.
Published/Produced	New Jersey: World Scientific, [2022]
Description	xii, 501 pages: color illustrations; 25 cm
ISBN	9789811231117 (hardcover)
	(ebook for institutions)
	(ebook for individuals)
LC classification	QC16.L56 A6 2022
Related names	Bartels, J. (Jochen), editor.
	Fadin, V. S. (Viktor Sergeevich), editor.
	Levin, Eugene (Eugene M.), editor.
	Levy, Aharon, 1940- editor.
	Kim, Victor T., editor.
	Sabio-Vera, Agustín, editor.
Summary	"This book has been designed to honor Lev Nikolaevich Lipatov, as a person and as one of

	the leading scientists in theoretical high energy physics. The book begins with three articles on Lev as a person, written endearingly by family members, a very close friend and Physics professor, Eugene Levin, and another outstanding scientist, Alfred Mueller. The book further collects 18 articles by several scientists who closely knew and/or collaborated with Lev. With an overarching range over various subfields, the book summarizes parts of Lev's achievements, presents new results which are based upon Lev's work, and paints an outlook on possible future developments. Lev's theoretical work has had an influential impact on phenomenology and experimental high energy physics; befittingly, this collection also includes several articles on these experimental aspects"--Provided by publisher.
Contents	Family memories: husband, father and grandfather / Elvira, Irina and Ekaterina Lipatov and grandchildren -- Lev Lipatov: my friend and renowned physicist / Eugene Levin -- Lev Lipatov: some personal reminiscences / A.H. Mueller -- BFKL - past and future / V.S. Fadin -- Basics of DGLAP / Yuri Dokshitzer -- Calculation of Green's functions in PT-symmetric quantum field theory / Carl M. Bender -- Rapidity evolution and small-x physics / Ian Balitsky -- Resummation at small x / Anna Maria Stasto -- From parton saturation to proton spin: the impact of BFKL equation and Reggeon evolution / Yuri V. Kovchegov -- The Odderon and BKP states in quantum chromodynamics / M.A. Braun and G.P. Vacca -- Lipatov's QCD

	high energy effective action: past and future / Martin Hentschinski -- High-energy scattering amplitudes in QED, QCD and supergravity / Agustín Sabio Vera -- N = 4 SYM quantum spectral curve in the BFKL regime / Mikhail Alfimov, Nikolay Gromov and Vladimir Kazakov -- DGLAP and BFKL equations in N = 4 SYM: from weak to strong coupling / A. Kotikov and A. Onishchenko -- The Regge limit of N = 4 SUSY Gauge theories / Jochen Bartels and Alex Prygarin -- The discrete BFKL pomeron and structure functions at low-x / H. Kowalski and D.A. Ross -- BFKL pomeron and the survival factor / V.A. Khoze, A.D. Martin and M.G. Ryskin -- A few topics in BFKL phenomenology at Hadron colliders / Agustín Sabio Vera -- Aspects of BFKL physics at HERA / H. Jung -- Aspects of BFKL physics at LEP, Tevatron and LHC / Victor T. Kim -- Lipatov's legacy and the future of deep inelastic scattering / Paul Newman.
Subjects	Lipatov, L. N. (Lev Nikolaevich), 1943-2017.
	Lipatov, L. N. (Lev Nikolaevich), 1943-2017--Influence.
	Physicists--Soviet Union--Biography.
	Physicists--Russia (Federation)--Biography.
	Particles (Nuclear physics)
Notes	Includes bibliographical references.
Additional formats	Online version: From the past to the future New Jersey: World Scientific, [2022] 9789811231124 (DLC) 2021049423

From the past to the future: the legacy of Lev Lipatov	
LCCN	2021049423
Type of material	Book
Main title	From the past to the future: the legacy of Lev Lipatov / editors, Joachim Bartels, Victor Fadin, Eugene Levin, Aharon Levy, Victor Kim, Agustín Sabio-Vera.
Published/Produced	New Jersey: World Scientific, [2022]
Description	1 online resource
ISBN	9789811231131 (ebook for individuals)
	9789811231124 (ebook for institutions)
	(hardcover)
LC classification	QC16.L56
Related names	Bartels, J. (Jochen), editor.
	Fadin, V. S. (Viktor Sergeevich), editor.
	Levin, Eugene (Eugene M.), editor.
	Levy, Aharon, 1940- editor.
	Kim, Victor T., editor.
	Sabio-Vera, Agustín, editor.
Summary	"This book has been designed to honor Lev Nikolaevich Lipatov, as a person and as one of the leading scientists in theoretical high energy physics. The book begins with three articles on Lev as a person, written endearingly by family members, a very close friend and Physics professor, Eugene Levin, and another outstanding scientist, Alfred Mueller. The book further collects 18 articles by several scientists who closely knew and/or collaborated with Lev. With an overarching range over various subfields, the book summarizes parts of Lev's achievements, presents new results which are based upon Lev's work, and paints an outlook on

	possible future developments. Lev's theoretical work has had an influential impact on phenomenology and experimental high energy physics; befittingly, this collection also includes several articles on these experimental aspects"--Provided by publisher.
Contents	Family memories: husband, father and grandfather / Elvira, Irina and Ekaterina Lipatov and grandchildren -- Lev Lipatov: my friend and renowned physicist / Eugene Levin -- Lev Lipatov: some personal reminiscences / A.H. Mueller -- BFKL - past and future / V.S. Fadin -- Basics of DGLAP / Yuri Dokshitzer -- Calculation of Green's functions in PT-symmetric quantum field theory / Carl M. Bender -- Rapidity evolution and small-x physics / Ian Balitsky -- Resummation at small x / Anna Maria Stasto -- From parton saturation to proton spin: the impact of BFKL equation and Reggeon evolution / Yuri V. Kovchegov -- The Odderon and BKP states in quantum chromodynamics / M.A. Braun and G.P. Vacca -- Lipatov's QCD high energy effective action: past and future / Martin Hentschinski -- High-energy scattering amplitudes in QED, QCD and supergravity / Agustín Sabio Vera -- N = 4 SYM quantum spectral curve in the BFKL regime / Mikhail Alfimov, Nikolay Gromov and Vladimir Kazakov -- DGLAP and BFKL equations in N = 4 SYM: from weak to strong coupling / A. Kotikov and A. Onishchenko -- The Regge limit of N = 4 SUSY Gauge theories / Jochen Bartels and Alex Prygarin -- The discrete BFKL pomeron and structure functions at low-x / H. Kowalski and D.A. Ross -- BFKL pomeron

	and the survival factor / V.A. Khoze, A.D. Martin and M.G. Ryskin -- A few topics in BFKL phenomenology at Hadron colliders / Agustín Sabio Vera -- Aspects of BFKL physics at HERA / H. Jung -- Aspects of BFKL physics at LEP, Tevatron and LHC / Victor T. Kim -- Lipatov's legacy and the future of deep inelastic scattering / Paul Newman.
Subjects	Lipatov, L. N. (Lev Nikolaevich), 1943-2017.
	Lipatov, L. N. (Lev Nikolaevich), 1943-2017--Influence.
	Physicists--Soviet Union--Biography.
	Physicists--Russia (Federation)--Biography.
	Particles (Nuclear physics)
Notes	Includes bibliographical references.
	Description based on print version record and CIP data provided by publisher; resource not viewed.
Additional formats	Print version: From the past to the future New Jersey: World Scientific, [2022] 9789811231117 (DLC) 2021049422

Functional methods and models in quantum field theory	
LCCN	2019049281
Type of material	Book
Personal name	Fried, H. M. (Herbert Martin), author.
Main title	Functional methods and models in quantum field theory / H. M. Fried.
Edition	Dover edition
Published/Produced	Mineola, New York: Dover Publications, Inc., 2020.
ISBN	9780486828596 (trade paperback)
LC classification	QC174.45 .F76 2020

Summary	"This volume presents a unified description of the major soluble and approximate models of relativistic quantum field theory. The first half offers a compact expression and derivation of functional methods. The second part addresses the models themselves, employing elegant functional techniques to describe nearly all the soluble and approximate models. The level of presentation is such that students familiar with conventional field theoretic arguments should make the transition to a functional description without difficulty. Topics addressed in Part I on functional methods include the generating functional and the S-matrix, construction of the generating functional, noncanonical (chiral) generalizations, and special topics in quantum electrodynamics. Part II's examination of model approximations covers perturbation expansions, soluble models, no-recoil methods, relativistic eikonal physics, and speculations at high energy. This edition features a new preface and appendix by H. M. Fried"-- Provided by publisher.
Contents	Functional methods. Introduction -- The generating functionality and the S-matrix -- Construction of the generating functional -- Noncanonical (chiral) generalizations -- Special topics in quantum electrodynamics -- Model approximations. Perturbation expansions -- Soluble models -- No-recoil methods -- Relativistic eikonal physics -- Speculations at high energy -- Appendix.
Subjects	Quantum field theory.
Notes	Originally published: Cambridge, Mass.: MIT Press, 1972.

Gauge field theory in natural geometric language: a revisitation of mathematical notions of quantum physics	
LCCN	2020945386
Type of material	Book
Personal name	Canarutto, Daniel, author.
Main title	Gauge field theory in natural geometric language: a revisitation of mathematical notions of quantum physics / Daniel Canarutto.
Edition	First edition.
Published/Produced	Oxford; New York: Oxford University Press, 2020.
Description	xxix, 331 pages: illustrations; 24 cm
ISBN	0198861494 hardback
	9780198861492 hardback
LC classification	QC793.3.G38 C37 2020
Summary	"Gauge Field theory in Natural Geometric Language addresses the need to clarify basic mathematical concepts at the crossroad between gravitation and quantum physics. Selected mathematical and theoretical topics are exposed within a brief, integrated approach that exploits standard and non-standard notions, as well as recent advances, in a natural geometric language in which the role of structure groups can be regarded as secondary even in the treatment of the gauge fields themselves. In proposing an original bridge between physics and mathematics, this text will appeal not only to mathematicians who wish to understand some of the basic ideas involved in quantum particle physics, but also to physicists who are not satisfied with the usual mathematical presentations of their field."- publisher

Contents	Part 1. Classical geometry -- Bundle prolongations and connections -- Special algebraic notions -- Spinors and Minkowski space -- Spinor bundles and spacetime geometry -- Part 2. Pre-quantum field theory -- Classical gauge field theory -- Gauge field theory and gravitation -- Optical geometry -- Electroweak geometry and fields -- First-order theory of fields with arbitrary spin -- Infinitesimal deformations of ECD fields -- Part 3. Quantum geometry -- Generalised maps -- Special generalised densities on Minkowski spacetime -- Multi-particle spaces -- Bundles of quantum states -- Quantum bundles -- Part 3. Quantum fields -- Quantum fields -- Detectors -- Free quantum fields -- Electroweak extensions -- Basic notions in particle physics -- Scattering matrix computations -- Quantum electrodynamics -- On gauge freedom and interactions.
Subjects	Gauge fields (Physics)
	Particles (Nuclear physics)
	Gauge fields (Physics)
	Particles (Nuclear physics)
	Champs de jauge (physique)
	Particules (physique nucléaire)
Notes	Includes bibliographical references and index.

Gauge integral structures for stochastic calculus and quantum electrodynamics	
LCCN	2020016334
Type of material	Book
Personal name	Muldowney, P. (Patrick), 1946- author.

Main title	Gauge integral structures for stochastic calculus and quantum electrodynamics / Patrick Muldowney.
Published/Produced	Hoboken, NJ: Wiley, [2020]
Description	1 online resource
ISBN	9781119595526 (epub)
	9781119595502 (adobe pdf)
	(cloth)
LC classification	QA274.2
Contents	Stochastic integration -- Random variation -- Integration and probability -- Stochastic processes -- Brownian motion -- Stochastic sums -- Gauges for product spaces -- Quantum field theory -- Quantum electrodynamics.
Subjects	Stochastic analysis.
	Henstock-Kurzweil integral.
	Feynman integrals.
	Quantum electrodynamics--Mathematics.
Notes	Includes bibliographical references and index.
	Description based on print version record and CIP data provided by publisher; resource not viewed.
Additional formats	Print version: Muldowney, P. (Patrick), 1946- Gauge integral structures for stochastic calculus and quantum electrodynamics Hoboken, NJ: Wiley, [2020] 9781119595496 (DLC) 2020016333

Gauge integral structures for stochastic calculus and quantum electrodynamics	
LCCN	2020016333
Type of material	Book
Personal name	Muldowney, P. (Patrick), 1946- author.

Main title	Gauge integral structures for stochastic calculus and quantum electrodynamics / Patrick Muldowney.
Published/Produced	Hoboken, NJ: Wiley, [2020]
ISBN	9781119595496 (cloth)
	(adobe pdf)
	(epub)
LC classification	QA274.2 .M85 2020
Contents	Stochastic integration -- Random variation -- Integration and probability -- Stochastic processes -- Brownian motion -- Stochastic sums -- Gauges for product spaces -- Quantum field theory -- Quantum electrodynamics.
Subjects	Stochastic analysis.
	Henstock-Kurzweil integral.
	Feynman integrals.
	Quantum electrodynamics--Mathematics.
Notes	Includes bibliographical references and index.
Additional formats	Online version: Muldowney, P. (Patrick), 1946- Gauge integral structures for stochastic calculus and quantum electrodynamics Hoboken, NJ: Wiley, [2020] 9781119595502 (DLC) 2020016334

GRIBOV-90 MEMORIAL: field theory, symmetry and related topics	
LCCN	2021425362
Type of material	Book
Main title	GRIBOV-90 MEMORIAL: field theory, symmetry, and related topics.
Published/Created	[S.l.]: WORLD SCIENTIFIC PUB, 2021.
Description	x, 515 pages: illustrations (black and white); 24 cm

ISBN	9811238391
	9789811238390
	ebook for institutions
	ebook for individuals
Subjects	Quantum field theory--Congresses.
	Particles (Nuclear physics)--Congresses.
Notes	Includes bibliographical references.

Handbook of the Tutte polynomial and related topics	
LCCN	2021048726
Type of material	Book
Main title	Handbook of the Tutte polynomial and related topics / edited by Joanna A. Ellis-Monaghan, Iain Moffatt.
Edition	First edition.
Published/Produced	Boca Raton: C&H/CRC Press, 2022.
Description	1 online resource
ISBN	9780429161612 (ebook)
	(hardback)
	(paperback)
LC classification	QA166.249
Related names	Ellis-Monaghan, Joanna Anthony, editor.
	Moffatt, Iain, editor.
Summary	"The Tutte Polynomial touches on nearly every area of combinatorics as well as many other fields, including statistical mechanics, coding theory, and DNA sequencing. It is one of the most studied graph polynomials. Handbook of the Tutte Polynomial and Related Topics is the first handbook published on the Tutte Polynomial. It consists of thirty-four chapters, written by experts in the field, that collectively offer a concise overview of the polynomial's

	many properties and applications. Features: Written in an accessible style for non-experts yet extensive enough for experts; Serves as a comprehensive and accessible introduction to the theory of graph polynomials for researchers in mathematics, physics, and computer science; Provides an extensive reference volume for the evaluations, theorems, and properties of the Tutte polynomial and related graph, matroid, and knot invariants; Offers broad coverage, touching on the wide range of applications of the Tutte polynomial and its various specializations"-- Provided by publisher.
Contents	Graph theory / Joanna A. Ellis-Monaghan, Iain Moffatt -- The Tutte polynomial for graphs / Joanna A. Ellis-Monaghan, Iain Moffatt -- Essential properties of the Tutte polynomial / Béla Bollobás, Oliver Riordan -- Matroid theory / James Oxley -- Tutte polynomial activities / Spencer Backman -- Tutte uniqueness and Tutte equivalence / Joseph E. Bonin, Anna de Mier -- Computational techniques / Criel Merino -- Computational resources / David Pearce, Gordon F. Royle -- The exact complexity of the Tutte polynomial / Tomer Kotek, Johann A. Makowsky -- Approximating the Tutte polynomial / Magnus Bordewich -- Foundations of the chromatic polynomial / Fengming Dong, Khee Meng Koh -- Flows and colorings / Delia Garijo, Andrew Goodall, Jaroslav Nešetřil -- Skein polynomials and the Tutte polynomial when x = y / Joanna A. Ellis-Monaghan, Iain Moffatt -- The interlace polynomial and the Tutte-Martin polynomial / Robert Brijder,

	Hendrik Jan Hoogeboom -- Network reliability / Jason I. Brown, Charles J. Colbourn -- Codes / Thomas Britz, Peter J. Cameron -- The chip-firing game and the sandpile model / Criel Merino -- The Tutte polynomial and knot theory / Stephen Huggett -- Quantum field theory connections / Adrian Tanasa -- The Potts and random-cluster models / Geoffrey Grimmett -- Where Tutte and Holant meet: a view from counting complexity / Jin-Yi Cai, Tyson Williams -- Polynomials and graph homomorphisms / Delia Garijo, Andrew Goodall, Jaroslav Nešetřil, Guus Regts -- Digraph analogues of the Tutte polynomial / Timothy Y. Chow -- Multivariable, parameterized, and colored extensions of the Tutte polynomial / Lorenzo Traldi -- Zeros of the Tutte polynomial / Bill Jackson -- The U, V and W polynomials / Steven Noble -- Topological extensions of the Tutte polynomial / Sergei Chmutov -- The Tutte polynomial of matroid perspectives / Emeric Gioan -- Hyperplane arrangements and the finite field method / Federico Ardila -- Some algebraic structures related to the Tutte polynomial / Michael J. Falk, Joseph P.S. Kung -- The Tutte polynomial of oriented matroids / Emeric Gioan -- Valuative invariants on matroid basis polytopes / Michael J. Falk, Joseph P.S. Kung -- Non-matroidal generalizations / Gary Gordon, Elizabeth McMahon -- The history of Tutte-Whitney polynomials / Graham Farr.
Subjects	Tutte polynomial.
	Graph theory.

	Polynomials.
	Invariants.
Notes	Includes bibliographical references and index.
	Description based on print version record and CIP data provided by publisher; resource not viewed.
Additional formats	Print version: Handbook of the Tutte polynomial and related topics First edition. Boca Raton: C&H/CRC Press, 2022 9781482240627 (DLC) 2021048725

Handbook of the Tutte polynomial and related topics	
LCCN	2021048725
Type of material	Book
Main title	Handbook of the Tutte polynomial and related topics / edited by Joanna A. Ellis-Monaghan, Iain Moffatt.
Edition	First edition.
Published/Produced	Boca Raton: C&H/CRC Press, 2022.
ISBN	9781482240627 (hardback)
	9781032228945 (paperback)
	(ebook)
LC classification	QA166.249 .H36 2022
Related names	Ellis-Monaghan, Joanna Anthony, editor.
	Moffatt, Iain, editor.
Summary	"The Tutte Polynomial touches on nearly every area of combinatorics as well as many other fields, including statistical mechanics, coding theory, and DNA sequencing. It is one of the most studied graph polynomials. Handbook of the Tutte Polynomial and Related Topics is the first handbook published on the Tutte Polynomial. It consists of thirty-four chapters,

	written by experts in the field, that collectively offer a concise overview of the polynomial's many properties and applications. Features: Written in an accessible style for non-experts yet extensive enough for experts; Serves as a comprehensive and accessible introduction to the theory of graph polynomials for researchers in mathematics, physics, and computer science; Provides an extensive reference volume for the evaluations, theorems, and properties of the Tutte polynomial and related graph, matroid, and knot invariants; Offers broad coverage, touching on the wide range of applications of the Tutte polynomial and its various specializations"-- Provided by publisher.
Contents	Graph theory / Joanna A. Ellis-Monaghan, Iain Moffatt -- The Tutte polynomial for graphs / Joanna A. Ellis-Monaghan, Iain Moffatt -- Essential properties of the Tutte polynomial / Béla Bollobás, Oliver Riordan -- Matroid theory / James Oxley -- Tutte polynomial activities / Spencer Backman -- Tutte uniqueness and Tutte equivalence / Joseph E. Bonin, Anna de Mier -- Computational techniques / Criel Merino -- Computational resources / David Pearce, Gordon F. Royle -- The exact complexity of the Tutte polynomial / Tomer Kotek, Johann A. Makowsky -- Approximating the Tutte polynomial / Magnus Bordewich -- Foundations of the chromatic polynomial / Fengming Dong, Khee Meng Koh -- Flows and colorings / Delia Garijo, Andrew Goodall, Jaroslav Nešetřil -- Skein polynomials and the Tutte polynomial when x = y / Joanna A. Ellis-Monaghan, Iain

	Moffatt -- The interlace polynomial and the Tutte-Martin polynomial / Robert Brijder, Hendrik Jan Hoogeboom -- Network reliability / Jason I. Brown, Charles J. Colbourn -- Codes / Thomas Britz, Peter J. Cameron -- The chip-firing game and the sandpile model / Criel Merino -- The Tutte polynomial and knot theory / Stephen Huggett -- Quantum field theory connections / Adrian Tanasa -- The Potts and random-cluster models / Geoffrey Grimmett -- Where Tutte and Holant meet: a view from counting complexity / Jin-Yi Cai, Tyson Williams -- Polynomials and graph homomorphisms / Delia Garijo, Andrew Goodall, Jaroslav Nešetřil, Guus Regts -- Digraph analogues of the Tutte polynomial / Timothy Y. Chow -- Multivariable, parameterized, and colored extensions of the Tutte polynomial / Lorenzo Traldi -- Zeros of the Tutte polynomial / Bill Jackson -- The U, V and W polynomials / Steven Noble -- Topological extensions of the Tutte polynomial / Sergei Chmutov -- The Tutte polynomial of matroid perspectives / Emeric Gioan -- Hyperplane arrangements and the finite field method / Federico Ardila -- Some algebraic structures related to the Tutte polynomial / Michael J. Falk, Joseph P.S. Kung -- The Tutte polynomial of oriented matroids / Emeric Gioan -- Valuative invariants on matroid basis polytopes / Michael J. Falk, Joseph P.S. Kung -- Non-matroidal generalizations / Gary Gordon, Elizabeth McMahon -- The history of Tutte-Whitney polynomials / Graham Farr.

Subjects	Tutte polynomial.
	Graph theory.
	Polynomials.
	Invariants.
Notes	Includes bibliographical references and index.
Additional formats	Online version: Handbook of the Tutte polynomial and related topics. First edition Boca Raton: C&H/CRC Press, 2022 9780429161612 (DLC) 2021048726

Higher spin field theory. Volume 1, Free theory	
LCCN	2020934839
Type of material	Book
Personal name	Bengtsson, Anders, author.
Main title	Higher spin field theory. Volume 1, Free theory / Anders Bengtsson.
Published/Produced	Berlin; Boston: De Gruyter, [2020]
Description	xviii, 354 pages: illustrations (black and white); 25 cm
ISBN	9783110450538 (hardcover)
	3110450534 (hardcover)
LC classification	QC793.3.G38 B46 2020
Summary	This monograph takes stock of the situation in higher spin gauge theories for the first time. Besides a thorough recapitulation of the field's history, it reviews the progress that has been made and offers a pedagogical introduction to the subject. Abstract approaches to the theory are offered to facilitate a conceptual rethinking of the main problems and to help see patterns hidden by heavy.-- Source other than the Library of Congress.

Contents	Frontmatter -- Preface -- Contents -- 1. Introduction and motivation -- 2. Notes on the history of the subject -- 3. Concepts, mathematical structures and notation -- 4. Lower spin theory -- 5. Exploring the free field theory -- 6. The light-front approach -- A. Epilogue -- Bibliography -- Index
Subjects	Gauge fields (Physics)
	Quantum field theory.
	Particles (Nuclear physics)
	Gauge fields (Physics)
	Particles (Nuclear physics)
	Quantum field theory.
Notes	Includes bibliographical references and index.
Additional formats	Electronic version: Bengtsson, Anders. Higher spin field theory. Volume 1, Free theory. Berlin; Boston: De Gruyter, [2020] 9783110451771 (DLC) 2020934839 (OCoLC)1163877965
Series	Texts and Monographs in Theoretical Physics, 2627-3934
	Texts and monographs in theoretical physics. 2627-3934

Homotopical quantum field theory	
LCCN	2019049784
Type of material	Book
Personal name	Yau, Donald Y. (Donald Ying), 1977- author.
Main title	Homotopical quantum field theory / Donald Yau.
Published/Produced	Hackensack, New Jersey: World Scientific, [2020]
ISBN	9789811212857 (hardcover)
	(ebook)
LC classification	QA612.7 .Y38 2020

Summary	"This book provides a general and powerful definition of homotopy algebraic quantum field theory and homotopy prefactorization algebra using a new coend definition of the Boardman-Vogt construction for a colored operad. All of their homotopy coherent structures are explained in details, along with a comparison between the two approaches at the operad level. With chapters on basic category theory, trees, and operads, this book is self-contained and is accessible to graduate students"-- Provided by publisher.
Contents	Category theory -- Trees -- Colored operads -- Constructions on operads -- Boardman-Vogt construction of operads -- Algebras over the Boardman-Vogt construction -- Algebraic quantum field theories -- Homotopy algebraic quantum field theories -- Prefactorization algebras -- Homotopy prefactorization algebras -- Comparing prefactorization algebras and AQFT.
Subjects	Homotopy theory.
	Quantum field theory--Mathematics.
Notes	Includes bibliographical references and index.

Homotopical quantum field theory	
LCCN	2019723536
Type of material	Book
Personal name	Yau, Donald Y. (Donald Ying), 1977- author.
Main title	Homotopical quantum field theory / Donald Yau.
Published/Produced	Hackensack, New Jersey: World Scientific, [2020]
Description	1 online resource

ISBN	9789811212864 (ebook)
	(hardcover)
LC classification	QA612.7
Summary	"This book provides a general and powerful definition of homotopy algebraic quantum field theory and homotopy prefactorization algebra using a new coend definition of the Boardman-Vogt construction for a colored operad. All of their homotopy coherent structures are explained in details, along with a comparison between the two approaches at the operad level. With chapters on basic category theory, trees, and operads, this book is self-contained and is accessible to graduate students"-- Provided by publisher.
Contents	Category theory -- Trees -- Colored operads -- Constructions on operads -- Boardman-Vogt construction of operads -- Algebras over the Boardman-Vogt construction -- Algebraic quantum field theories -- Homotopy algebraic quantum field theories -- Prefactorization algebras -- Homotopy prefactorization algebras -- Comparing prefactorization algebras and AQFT.
Subjects	Homotopy theory.
	Quantum field theory--Mathematics.
Notes	Includes bibliographical references and index.
	Description based on print version record.
Additional formats	Print version: Homotopical quantum field theory Hackensack, New Jersey: World Scientific, [2020] 9789811212857 (DLC) 2019049784

Integrability, quantization, and geometry	
LCCN	2020043148
Type of material	Book
Main title	Integrability, quantization, and geometry / Igor Krichever, Sergey Novikov, Oleg Ogievetsky, Senya Shlosman, editors.
Published/Produced	Providence, Rhode Island: American Mathematical Society, [2021]
Description	2 volumes (xxi, 516; xxi, 480 pages): illustrations, portraits; 26 cm
ISBN	9781470455910 (volume 1; paperback)
	9781470455927 (volume 2; paperback)
	(volume 1; ebook)
	(volume 2; ebook)
LC classification	QA564 .I54 2021
Related names	Krichever, I. M., editor.
Contents	Volume 1. Integrable systems -- Volume 2. Quantum theories and algebraic geometry.
Subjects	Dubrovin, B. A. (Boris Anatol'evich)--Influence.
	Geometry, Algebraic.
	Topology.
	Homology theory.
	Quantum theory.
	Algebraic geometry -- (Co)homology theory [See also 13Dxx] -- Sheaves, derived categories of sheaves and related constructions [See also 14H60, 14J60, 18F20, 32Lxx, 46M20].
	Algebraic geometry -- (Co)homology theory [See also 13Dxx] -- Differentials and other special sheaves; D-modules; Bernstein-Sato ideals and polynomials [See also 13Nxx, 32C38].

	Algebraic geometry -- Arithmetic problems. Diophantine geometry [See also 11Dxx, 11Gxx] -- Zeta-functions and related questions [See also 11G40] (Birch-Swinnerton-Dyer conjecture).
	Algebraic geometry -- Surfaces and higher-dimensional varieties {For analytic theory, see 32Jxx} -- Families, moduli, classification: algebraic theory.
	Algebraic geometry -- Surfaces and higher-dimensional varieties {For analytic theory, see 32Jxx} -- Mirror symmetry [See also 11G42, 53D37].
	Algebraic geometry -- Surfaces and higher-dimensional varieties {For analytic theory, see 32Jxx} -- Vector bundles on surfaces and higher-dimensional varieties, and their moduli [See also 14D20, 14F05
	Special functions (33-XX deals with the properties of functions as functions) {For orthogonal functions, see 42Cxx; for aspects of combinatorics see 05Axx; for number-theoretic aspects see 11-XX; for
	Differential geometry {For differential topology, see 57Rxx. For foundational questions of differentiable manifolds, see 58Axx} -- Classical differential geometry -- Higher-dimensional and -codimensional
	Differential geometry {For differential topology, see 57Rxx. For foundational questions of differentiable manifolds, see 58Axx} -- Symplectic geometry, contact geometry [See also 37Jxx, 70Gxx, 70Hxx]
	Differential geometry {For differential topology, see 57Rxx. For foundational questions of

	differentiable manifolds, see 58Axx} -- Symplectic geometry, contact geometry [See also 37Jxx, 70Gxx, 70Hxx]
	Algebraic geometry -- Surfaces and higher-dimensional varieties {For analytic theory, see 32Jxx} -- Relationships with physics.
	Associative rings and algebras {For the commutative case, see 13-XX} -- Hopf algebras, quantum groups and related topics -- Ring-theoretic aspects of quantum groups [See also 17B37, 20G42, 81R50].
	Group theory and generalizations -- Representation theory of groups [See also 19A22 (for representation rings and Burnside rings)] -- Ordinary representations and characters.
	Group theory and generalizations -- Representation theory of groups [See also 19A22 (for representation rings and Burnside rings)] -- Applications of group representations to physics.
	Convex and discrete geometry -- Discrete geometry -- Lattices and convex bodies in n dimensions [See also 11H06, 11H31, 11P21].
	Differential geometry {For differential topology, see 57Rxx. For foundational questions of differentiable manifolds, see 58Axx} -- Symplectic geometry, contact geometry [See also 37Jxx, 70Gxx, 70Hxx]
	Differential geometry {For differential topology, see 57Rxx. For foundational questions of differentiable manifolds, see 58Axx} -- Symplectic geometry, contact geometry [See also 37Jxx, 70Gxx, 70Hxx]
	Quantum theory -- Quantum field theory; related classical field theories [See also 70Sxx] -- Yang-

	Mills and other gauge theories [See also 53C07, 58E15].
	Quantum theory -- Quantum field theory; related classical field theories [See also 70Sxx] -- String and superstring theories; other extended objects (e.g., branes) [See also 83E30].
	Quantum theory -- Quantum field theory; related classical field theories [See also 70Sxx] -- Two-dimensional field theories, conformal field theories, etc..
Notes	Includes bibliographical references.
Series	Proceedings of symposia in pure mathematics, 0082-0717; 103.1, 103.2

Introduction to quantum field theory and the standard model	
LCCN	2021050560
Type of material	Book
Personal name	Hollik, W. (Wolfgang), 1951- author.
Main title	Introduction to quantum field theory and the standard model / Wolfgang Hollik, Max Planck Institute for Physics, Germany.
Published/Produced	New Jersey: World Scientific, [2022]
Description	1 online resource
ISBN	9789811242199 (ebook other)
	9789811242182 (ebook)
	(hardcover)
LC classification	QC174.45
Summary	"Based on the lectures given at TU Munich for third-year physics students, this book provides the basic concepts of relativistic quantum field theory, perturbation theory, Feynman graphs, Abelian and non-Abelian gauge theories, with application to QED, QCD, and the electroweak

	Standard Model. It also introduces quantum field theory and particle physics for beginning graduate students with an orientation towards particle physics and its theoretical foundations. Phenomenology of W and Z bosons, as well as Higgs bosons, is part of the electroweak chapter in addition to recent experimental results, precision tests and current status of the Standard Model"-- Provided by publisher.
Subjects	Quantum field theory.
Notes	Includes bibliographical references and index.
	Description based on print version record and CIP data provided by publisher; resource not viewed.
Additional formats	Print version: Hollik, W. 1951- Introduction to quantum field theory and the standard model New Jersey: World Scientific, [2022] 9789811242175 (DLC) 2021050559

Introduction to quantum field theory and the standard model	
LCCN	2021050559
Type of material	Book
Personal name	Hollik, W. (Wolfgang), 1951- author.
Main title	Introduction to quantum field theory and the standard model / Wolfgang Hollik, Max Planck Institute for Physics, Germany.
Published/Produced	New Jersey: World Scientific, [2022]
ISBN	9789811242175 (hardcover)
	(ebook)
	(ebook other)
LC classification	QC174.45 .H647 2022
Summary	"Based on the lectures given at TU Munich for third-year physics students, this book provides

	the basic concepts of relativistic quantum field theory, perturbation theory, Feynman graphs, Abelian and non-Abelian gauge theories, with application to QED, QCD, and the electroweak Standard Model. It also introduces quantum field theory and particle physics for beginning graduate students with an orientation towards particle physics and its theoretical foundations. Phenomenology of W and Z bosons, as well as Higgs bosons, is part of the electroweak chapter in addition to recent experimental results, precision tests and current status of the Standard Model"-- Provided by publisher.
Subjects	Quantum field theory.
Notes	Includes bibliographical references and index.
Additional formats	Online version: Hollik, Wolfgang. Introduction to quantum field theory and the standard model Hackensack: World Scientific, 2022 9789811242182 (DLC) 2021050560

Introduction to quantum field theory with applications to quantum gravity	
LCCN	2020945719
Type of material	Book
Personal name	Buchbinder, Joseph, author.
Main title	Introduction to quantum field theory with applications to quantum gravity / Iosif L. Buchbinder, Ilya L. Shapiro.
Edition	First edition.
Published/Produced	Oxford: Oxford University Press, 2021.
Description	x, 525 pages: illustrations (black and white); 26 cm.
ISBN	9780198838319 hardback

	019883831X hardback
LC classification	QC174.45 .B83 2021
Related names	Shapiro, Ilya., author.
Subjects	Quantum field theory.
	Quantum gravity.
	Quantum field theory.
	Quantum gravity.
Notes	Includes bibliographical references and index.
Series	Oxford graduate texts
	Oxford graduate texts.

Introduction to quantum field theory	
LCCN	2019006491
Type of material	Book
Personal name	Năstase, Horațiu, 1972- author.
Main title	Introduction to quantum field theory / Horațiu Năstase (Universidade Estadual Paulista, São Paulo).
Published/Produced	Cambridge, United Kingdom; New York, NY: Cambridge University Press, 2020.
	©2020
ISBN	9781108493994 (alk. paper)
	1108493998 (alk. paper)
LC classification	QC174.45 .N353 2020
Summary	"This book is meant as a two-semester course in quantum field theory, skipping some material that can be studied independently. The chapters with asterisk I have not taught in my class, and can be skipped in a first reading, or when teaching the material. The book and the corresponding course is supposed to follow a course in classical field theory, however, I have tried to make the book self-contained. That

	means that only a thorough knowledge of classical mechanics, quantum mechanics, and electromagnetism is really needed, though it is preferrable to have firrst classical field theory. I will only review classical field theory, without going in great detail"-- Provided by publisher.
Subjects	Quantum field theory.
Notes	Includes bibliographical references and index.

Introduction to the standard model and beyond: quantum field theory, symmetries and phenomenology	
LCCN	2021024621
Type of material	Book
Personal name	Raby, Stuart, author.
Main title	Introduction to the standard model and beyond: quantum field theory, symmetries and phenomenology / Stuart Raby.
Published/Produced	Cambridge: Cambridge University Press, 2021.
Description	1 online resource
ISBN	9781108644129 (ebook)
	(hardback)
LC classification	QC794.6.S75
Summary	"The Standard Model of particle physics is an amazingly successful theory describing the fundamental particles and forces of nature. This text, written for a two-semester graduate course on the Standard Model, develops a practical understanding of the theoretical concepts it's built upon, to prepare students to enter research. The author takes a historical approach to demonstrate to students the process of discovery which is often overlooked in other textbooks, presenting quantum field theory and symmetries as the

	necessary tools for describing and understanding the Standard Model. He develops these tools using a basic understanding of quantum mechanics and classical field theory, such as Maxwell's electrodynamics, before discussing the important role that Noether's theorem and conserved charges play in the theory. Worked examples feature throughout the text, while homework exercises are included for the first five parts, with solutions available online for instructors. Inspired by the author's own teaching experience, suggestions for independent research topics have been provided for the second-half of the course, which students can then present to the rest of the class"-- Provided by publisher.
Subjects	Standard model (Nuclear physics)
	Quantum theory.
	Science / Physics / Nuclear
	Science / Physics / Nuclear
Notes	Includes bibliographical references and index.
	Description based on print version record and CIP data provided by publisher; resource not viewed.
Additional formats	Print version: Raby, Stuart. Introduction to the standard model and beyond Cambridge: Cambridge University Press, 2021 9781108494199 (DLC) 2021024620

Introduction to the standard model and beyond: quantum field theory, symmetries and phenomenology	
LCCN	2021024620
Type of material	Book
Personal name	Raby, Stuart, author.

Main title	Introduction to the standard model and beyond: quantum field theory, symmetries and phenomenology / Stuart Raby.
Published/Produced	Cambridge: Cambridge University Press, 2021.
ISBN	9781108494199 (hardback)
	(ebook)
LC classification	QC794.6.S75 R33 2021
Summary	"The Standard Model of particle physics is an amazingly successful theory describing the fundamental particles and forces of nature. This text, written for a two-semester graduate course on the Standard Model, develops a practical understanding of the theoretical concepts it's built upon, to prepare students to enter research. The author takes a historical approach to demonstrate to students the process of discovery which is often overlooked in other textbooks, presenting quantum field theory and symmetries as the necessary tools for describing and understanding the Standard Model. He develops these tools using a basic understanding of quantum mechanics and classical field theory, such as Maxwell's electrodynamics, before discussing the important role that Noether's theorem and conserved charges play in the theory. Worked examples feature throughout the text, while homework exercises are included for the first five parts, with solutions available online for instructors. Inspired by the author's own teaching experience, suggestions for independent research topics have been provided for the second-half of the course, which students can then present to the rest of the class"--Provided by publisher.

Subjects	Standard model (Nuclear physics)
	Quantum theory.
	Science / Physics / Nuclear
	Science / Physics / Nuclear
Notes	Includes bibliographical references and index.
Additional formats	Online version: Raby, Stuart Introduction to the standard model and beyond Cambridge: Cambridge University Press, 2021 9781108644129 (DLC) 2021024621

Lattice quantum field theory of the Dirac and gauge fields: selected topics	
LCCN	2019047773
Type of material	Book
Personal name	Baaquie, B. E., author.
Main title	Lattice quantum field theory of the Dirac and gauge fields: selected topics / Belal Ehsan Baaquie.
Published/Produced	New Jersey: World Scientific Publishing, [2020]
ISBN	9789811209697 (hardcover)
	(ebook for institutions)
	(ebook for individuals)
LC classification	QC793.3.G38 B33 2020
Summary	"Quantum Chromodynamics is the theory of strong interactions: a quantum field theory of colored gluons (Yang-Mills gauge fields) coupled to quarks (Dirac fermion fields). Lattice gauge theory is defined by discretizing spacetime into a four-dimensional lattice --- and entails defining gauge fields and Dirac fermions on a lattice. The applications of lattice gauge theory are vast, from the study of high-energy theory and phenomenology to the numerical studies of

	quantum fields. "Lattice Quantum Field Theory of the Dirac and Gauge Fields: Selected Topics" examines the mathematical foundations of lattice gauge theory from first principles. It is indispensable for the study of Dirac and lattice gauge fields and lays the foundation for more advanced and specialized studies"-- Provided by publisher.
Contents	SU(N) compact lie groups -- SU(N) Kac-Moody algebra -- SU(N) path integrals -- SU(3) character functions -- Fermion calculus -- Non-Abelian lattice gauge field -- Abelian lattice gauge field in d=3 -- Lattice gauge field mass renormalization -- Gauge field block-spin renormalization -- Lattice gauge field Hamiltonian -- Dirac lattice path integral -- Dirac Hamiltonian -- Lattice gauge theory Hamiltonian.
Subjects	Lattice gauge theories.
	Quantum field theory.
	Lattice field theory.
	Dirac equation.
	Quantum chromodynamics.
Notes	Includes bibliographical references and index.

Lattice quantum field theory of the Dirac and gauge fields: selected topics	
LCCN	2019723406
Type of material	Book
Personal name	Baaquie, B. E., author.
Main title	Lattice quantum field theory of the Dirac and gauge fields: selected topics / Belal Ehsan Baaquie.

Published/Produced	New Jersey: World Scientific Publishing, [2020]
Description	1 online resource
ISBN	9789811209703 (ebook for institutions)
	9789811209710 (ebook for individuals)
	(hardcover)
LC classification	QC793.3.G38
Summary	"Quantum Chromodynamics is the theory of strong interactions: a quantum field theory of colored gluons (Yang-Mills gauge fields) coupled to quarks (Dirac fermion fields). Lattice gauge theory is defined by discretizing spacetime into a four-dimensional lattice --- and entails defining gauge fields and Dirac fermions on a lattice. The applications of lattice gauge theory are vast, from the study of high-energy theory and phenomenology to the numerical studies of quantum fields. "Lattice Quantum Field Theory of the Dirac and Gauge Fields: Selected Topics" examines the mathematical foundations of lattice gauge theory from first principles. It is indispensable for the study of Dirac and lattice gauge fields and lays the foundation for more advanced and specialized studies"-- Provided by publisher.
Contents	SU(N) compact lie groups -- SU(N) Kac-Moody algebra -- SU(N) path integrals -- SU(3) character functions -- Fermion calculus -- Non-Abelian lattice gauge field -- Abelian lattice gauge field in d=3 -- Lattice gauge field mass renormalization -- Gauge field block-spin renormalization -- Lattice gauge field Hamiltonian -- Dirac lattice path integral -- Dirac Hamiltonian -- Lattice gauge theory Hamiltonian.
Subjects	Lattice gauge theories.

	Quantum field theory.
	Lattice field theory.
	Dirac equation.
	Quantum chromodynamics.
Notes	Includes bibliographical references and index.
	Description based on print version record.
Additional formats	Print version: Lattice quantum field theory of the Dirac and gauge fields New Jersey: World Scientific Publishing, [2020] 9789811209697 (DLC) 2019047773

Lectures on quantum field theory	
LCCN	2020030278
Type of material	Book
Personal name	Das, Ashok, 1953- author.
Main title	Lectures on quantum field theory / Ashok Das, University of Rochester, USA.
Edition	Second edition.
Published/Produced	New Jersey: World Scientific, [2021]
ISBN	9789811220869 (hardcover)
	9789811222160 (paperback)
	(ebook for institutions)
	(ebook for individuals)
LC classification	QC174.45 .D369 2021
Portion of title	Quantum field theory
Summary	"This book comprises the lectures of a two-semester course on quantum field theory, presented in a quite informal and personal manner. The course starts with relativistic one-particle systems, and develops the basics of quantum field theory with an analysis on the representations of the Poincaré group. Canonical quantization is carried out for scalar, fermion,

	Abelian and non-Abelian gauge theories. Covariant quantization of gauge theories is also carried out with a detailed description of the BRST symmetry. The Higgs phenomenon and the standard model of electroweak interactions are also developed systematically. Regularization and (BPHZ) renormalization of field theories as well as gauge theories are discussed in detail, leading to a derivation of the renormalization group equation. In addition, two chapters - one on the Dirac quantization of constrained systems and another on discrete symmetries - are included for completeness, although these are not covered in the two-semester course. This second edition includes two new chapters, one on Nielsen identities and the other on basics of global supersymmetry. It also includes two appendices, one on fermions in arbitrary dimensions and the other on gauge invariant potentials and the Fock-Schwinger gauge"-- Provided by publisher.
Subjects	Quantum field theory--Textbooks.
Notes	Includes bibliographical references and index.

Local operators in integrable models I	
LCCN	2021006693
Type of material	Book
Personal name	Jimbo, M. (Michio), author.
Main title	Local operators in integrable models I / Michio Jimbo, Tetsuji Miwa, Fedor Smirnov.
Published/Produced	Providence, Rhode Island: American Mathematical Society, [2021]-
Description	xii, 192 pages: illustrations; 26 cm.
ISBN	9781470465520 (paperback; acid-free paper)

	(ebook)
LC classification	QC20.7.I58 J56 2021
Related names	Miwa, T. (Tetsuji), author.
	Smirnov, F. A., author.
Subjects	Integral equations.
	Operator theory.
	Quantum field theory.
	Statistical mechanics.
	Statistical mechanics, structure of matter -- Equilibrium statistical mechanics -- Lattice systems (Ising, dimer, Potts, etc.) and systems on graphs arising in equilibrium statistical mechanics.
	Statistical mechanics, structure of matter -- Equilibrium statistical mechanics -- Exactly solvable models; Bethe ansatz.
	Associative rings and algebras -- Hopf algebras, quantum groups and related topics -- Yang-Baxter equations.
	Nonassociative rings and algebras -- Lie algebras and Lie superalgebras -- Quantum groups (quantized enveloping algebras) and related deformations.
	Quantum theory -- Quantum field theory; related classical field theories -- Two-dimensional field theories, conformal field theories, etc. in quantum mechanics.
	Quantum theory -- Groups and algebras in quantum theory -- Quantum groups and related algebraic methods applied to problems in quantum theory.
Notes	Includes bibliographical references (pages 187-190) and index.

Series	Mathematical surveys and monographs; volume 256

Łojasiewicz-Simon gradient inequalities for coupled Yang-Mills energy functionals	
LCCN	2021016998
Type of material	Book
Personal name	Feehan, Paul M. N., 1961- author.
Main title	Łojasiewicz-Simon gradient inequalities for coupled Yang-Mills energy functionals / Paul M.N. Feehan, Manousos Maridakis.
Published/Produced	Providence: American Mathematical Society, [2020]
Description	xiii, 138 pages; 26 cm
ISBN	9781470443023 (paperback)
	(pdf)
LC classification	QC174.52.Y37 F44 2020
Related names	Maridakis, Manousos, 1982- author.
Summary	"We prove Lojasiewicz-Simon gradient inequalities for coupled Yang-Mills energy functions using Sobolev spaces which impose minimal regularity requirements on pairs of connections and sections. The Lojasiewicz-Simon gradient inequalities for coupled Yang-Mills energy functions generalize that of the pure Yang-Mills energy function due to the first author (Feehan, 2014) for base manifolds of arbitrary dimension and due to Räde (1992, Proposition 7.2) for dimensions two and three"--Provided by publisher.
Subjects	Yang-Mills theory.
	Global analysis, analysis on manifolds [See also 32Cxx, 32Fxx, 32Wxx, 46-XX, 47Hxx, 53Cxx]

	{For geometric integration theory, see 49Q15} -- Variational problems in infinite-dimensional spaces -- Applications
	Manifolds and cell complexes {For complex manifolds, see 32Qxx} -- Differential topology {For foundational questions of differentiable manifolds, see 58Axx; for infinite-dimensional manifolds, see 58B
	Dynamical systems and ergodic theory [See also 26A18, 28Dxx, 34Cxx, 34Dxx, 35Bxx, 46Lxx, 58Jxx, 70-XX] -- Dynamical systems with hyperbolic behavior -- Morse-Smale systems.
	Global analysis, analysis on manifolds [See also 32Cxx, 32Fxx, 32Wxx, 46-XX, 47Hxx, 53Cxx] {For geometric integration theory, see 49Q15} -- Spaces and manifolds of mappings (including nonlinear version
	Mechanics of particles and systems {For relativistic mechanics, see 83A05 and 83C10; for statistical mechanics, see 82-XX} -- Classical field theories [See also 37Kxx, 37Lxx, 78-XX, 81Txx, 83-XX] -- Y
	Quantum theory -- Quantum field theory; related classical field theories [See also 70Sxx] -- Yang-Mills and other gauge theories [See also 53C07, 58E15].
Notes	Includes bibliographical references.
Series	Memoirs of the American Mathematical Society, 0065-9266; Number 1302

Mathematical foundations of quantum field theory	
LCCN	2019034999
Type of material	Book

Personal name	Shvarts, A. S. (Al'bert Solomonovich), author.
Main title	Mathematical foundations of quantum field theory / Albert Schwarz, University of California at Davis.
Published/Produced	New Jersey: World Scientific, [2020]
ISBN	9789813278639 (hardcover)
LC classification	QC174.45 .S3295 2020
Summary	"The book is very different from other books devoted to quantum field theory, both in the style of exposition and in the choice of topics. Written for both mathematicians and physicists, the author explains the theoretical formulation with a mixture of rigorous proofs and heuristic arguments; references are given for those who are looking for more details. The author is also careful to avoid ambiguous definitions and statements that can be found in some physics textbooks. In terms of topics, almost all other books are devoted to relativistic quantum field theory, conversely this book is concentrated on the material that does not depend on the assumptions of Lorentz-invariance and/or locality. It contains also a chapter discussing application of methods of quantum field theory to statistical physics, in particular to the derivation of the diagram techniques that appear in thermo-field dynamics and Keldysh formalism. It is not assumed that the reader is familiar with quantum mechanics; the book contains a short introduction to quantum mechanics for mathematicians and an appendix devoted to some mathematical facts used in the book"-- Provided by publisher.
Subjects	Quantum field theory.

Notes	Includes bibliographical references.

Phenomenology of particle physics	
LCCN	2021044753
Type of material	Book
Personal name	Rubbia, André, 1966- author.
Main title	Phenomenology of particle physics / André Rubbia.
Published/Produced	Cambridge; New York, NY: Cambridge University Press, 2022.
ISBN	9781316519349 (hardback)
LC classification	QC793.29 .R83 2022
Summary	"Particle physics intertwines theory and experiments, and this text demonstrates and develops the interplay between the two, following the author's detailed and original approach. This complete and comprehensive treatise, written for a two-semester Master's or graduate course, covers all aspects of modern particle physics. Richly illustrated with more than 450 figures, this text guides students through all the intricacies of quantum mechanics and quantum field theory in an intuitive manner that few books achieve. Featuring rigorous step-by-step derivations and more than 100 end-of-chapter problems for additional practice, it ensures that students will not only understand the material but also be able to apply their knowledge. Containing up-to-date experimental material, including the discovery of the Higgs boson at CERN and of neutrino oscillations, this monumental volume also serves as a one-stop

	reference for particle physics researchers of all levels and specialties"-- Provided by publisher.
Subjects	Particles (Nuclear physics)
	Phenomenological theory (Physics)
	Quantum theory.
	Quantum field theory.
	Science / Physics / Nuclear
Notes	Includes bibliographical references and index.

Problems in quantum field theory	
LCCN	2021038669
Type of material	Book
Personal name	Gelis, François, 1972- author.
Main title	Problems in quantum field theory / François Gelis.
Published/Produced	Cambridge, United Kingdom; New York, NY: Cambridge University Press, 2021.
ISBN	9781108838801 (hardback)
	9781108972352 (paperback)
LC classification	QC174.45 .G452 2021
Summary	"This chapter is devoted to basic aspects of quantum field theory, ranging from the foundations to perturbation theory and renormalization, and is limited to the canonical formalism (functional methods are treated in Chapter 2) and to the traditional workflow (Lagrangian-Feynman rules-time-ordered products of fields-scattering amplitudes) for the calculation of scattering amplitudes (the spinor-helicity formalism and on-shell recursion are considered in Chapter 4). The problems of this chapter deal with questions in scalar field theory and"-- Provided by publisher.
Subjects	Quantum field theory.

	Science / Physics / Mathematical & Computational
Notes	Includes bibliographical references and index.

Proceedings of the Eighth Meeting on CPT and Lorentz Symmetry, Indiana University, Bloomington, USA, 12-16 May 2019	
LCCN	2020008815
Type of material	Book
Meeting name	Meeting on CPT and Lorentz Symmetry (8th: 2019: Indiana University, Bloomington), author.
Main title	Proceedings of the Eighth Meeting on CPT and Lorentz Symmetry, Indiana University, Bloomington, USA, 12-16 May 2019 / editor, Ralf Lehnert.
Published/Produced	Hackensack, New Jersey: World Scientific, [2020]
ISBN	9789811213977 (hardcover)
	(ebook for institutions)
	(ebook for individuals)
LC classification	QC793.3.V5 M44 2020
Related names	Lehnert, R. (Ralf), editor.
Summary	"This book contains the Proceedings of the Eighth Meeting on CPT and Lorentz Symmetry, held at Indiana University in Bloomington on May 12-16, 2019. The Meeting focused on tests of these fundamental symmetries and on related theoretical issues, including scenarios for possible violations. Topics covered at the meeting include experimental and observational searches for CPT and Lorentz violation involving: accelerators and colliders; astrophysical birefringence, dispersion, and anisotropy; atomic and molecular spectroscopy;

	cavities, oscillators, resonators; Cherenkov radiation; clock-comparison measurements; CMB polarimetry; cosmic rays; decays of atoms, nuclei, and particles; equivalence-principle tests with matter and antimatter; exotic atoms, muonium, positronium; gauge bosons, the Higgs boson; gravimetry; gravitational waves; high-energy astrophysical observations; hydrogen and antihydrogen spectroscopy; lasers, masers; matter-wave interferometry; meson and baryon properties; neutral-meson interferometry; neutrino mixing and propagation, neutrino-antineutrino oscillations; particle-antiparticle comparisons; photon and particle scattering; post-Newton gravity in the solar system and beyond; second- and third-generation particles; short-range gravity; sidereal and annual time variations, compass asymmetries; single-top and top pair production; space-based missions; spin-gravity couplings; spin precession; time-of-flight measurements; torsion and nonmetricity; trapped particles, ions, and atoms. The meeting also covered theoretical and phenomenological studies of CPT and Lorentz violation including: physical effects at the level of the Standard Model, General Relativity, and beyond; origins and mechanisms for violations; classical and quantum field theory, gravitation, particle physics, and strings; mathematical foundations, Finsler geometry"-- Provided by publisher.
Subjects	CP violation (Nuclear physics)--Congresses.
	Lorentz groups--Congresses.
Notes	Includes bibliographical references and index.

Quantum computing: physics, blockchains, and deep learning smart networks	
LCCN	2019053514
Type of material	Book
Personal name	Swan, Melanie, author.
Main title	Quantum computing: physics, blockchains, and deep learning smart networks / Melanie Swan, Purdue University, USA, Renato P. dos Santos, Lutheran University of Brazil, Brazil, Frank Witte, University College London, UK.
Published/Produced	New Jersey: World scientific, [2020]
ISBN	9781786348203 (hardcover)
	(ebook)
	(ebook)
LC classification	QA76.889 .S93 2020
Summary	"Quantum information and contemporary smart network domains are so large and complex as to be beyond the reach of current research approaches. Hence, new theories are needed for their understanding and control. Physics is implicated as smart networks are physical systems comprised of particle-many items interacting and reaching criticality and emergence across volumes of macroscopic and microscopic states. Methods are integrated from statistical physics, information theory, and computer science. Statistical neural field theory and the AdS/CFT correspondence are employed to derive a smart network field theory (SNFT) and a smart network quantum field theory (SNQFT) for the orchestration of smart network systems. Specifically, a smart network field theory (conventional or quantum) is a field

	theory for the organization of particle-many systems from a characterization, control, criticality, and novelty emergence perspective"--Provided by publisher.
Subjects	Quantum computing.
	Blockchains (Databases)
	Finance--Technological innovations.
	Quantum field theory.
Notes	Includes bibliographical references and index.
Series	Between science and economics, 2051-6304; vol 2

Quantum field theory: a diagrammatic approach	
LCCN	2020056957
Type of material	Book
Personal name	Kleiss, Ronald, author.
Main title	Quantum field theory: a diagrammatic approach / Ronald Kleiss.
Published/Produced	New York: Cambridge University Press, 2021.
Description	1 online resource
ISBN	9781108665209 (ebook)
	(hardback)
LC classification	QC174.45
Summary	"Let us first consider a general Hamiltonian of Z electrons moving around a nucleus that contains the Coulomb interaction, the spin-orbit coupling, and the hyperfine coupling. These are the three effects that determine the electronic structure of an atom. Here we should emphasize the important role of the separation of energy scales; that is to say, the typical energy scales of these three terms are quite different. Thanks to the separation of energy scales, we can analyze them one by one,

	which enables us to obtain a clear picture of the electron structure"-- Provided by publisher.
Subjects	Quantum field theory.
Notes	Includes bibliographical references and index.
	Description based on print version record and CIP data provided by publisher; resource not viewed.
Additional formats	Print version: Kleiss, Ronald. Quantum field theory New York: Cambridge University Press, 2021. 9781108486217 (DLC) 2020056956

Quantum field theory: a diagrammatic approach	
LCCN	2020056956
Type of material	Book
Personal name	Kleiss, Ronald, author.
Main title	Quantum field theory: a diagrammatic approach / Ronald Kleiss.
Published/Produced	New York: Cambridge University Press, 2021.
ISBN	9781108486217 (hardback)
	(ebook)
LC classification	QC174.45 .K58 2021
Summary	"Let us first consider a general Hamiltonian of Z electrons moving around a nucleus that contains the Coulomb interaction, the spin-orbit coupling, and the hyperfine coupling. These are the three effects that determine the electronic structure of an atom. Here we should emphasize the important role of the separation of energy scales; that is to say, the typical energy scales of these three terms are quite different. Thanks to the separation of energy scales, we can analyze them one by one, which enables us to obtain a clear

	picture of the electron structure"-- Provided by publisher.
Subjects	Quantum field theory.
Notes	Includes bibliographical references and index.
Additional formats	Online version: Kleiss, Ronald, Quantum field theory New York: Cambridge University Press, 2021. 9781108665209 (DLC) 2020056957

Quantum field theory: an integrated approach	
LCCN	2020044827
Type of material	Book
Personal name	Fradkin, Eduardo, author.
Main title	Quantum field theory: an integrated approach / Eduardo Fradkin.
Published/Produced	Princeton: Princeton University Press, [2021]
Description	1 online resource
ISBN	9780691189550 (pdf)
	(hardback)
LC classification	QC174.45
Summary	"Quantum field theory is the mathematical and conceptual framework that describes the physics of the very small, including subatomic particles and quasiparticles. It is used to address a range of problems across subfields, from high-energy physics and gravitation to statistical physics and condensed matter physics. Despite the breadth of its applications, however, the teaching of quantum field theory has historically been strongly oriented toward high-energy physics students, while others-particularly in condensed matter and statistical physics-are typically taught in a separate course, or take an alternate sequence in many-body and statistical physics. Author Eduardo Fradkin

	strongly believes that this separation is both artificial and detrimental to all groups' understanding of quantum field theory. This textbook, developed from a graduate course Fradkin has taught for decades at the University of Illinois, offers a new, "multicultural" approach to the subject that seeks to remedy this fragmentation. It covers both basic techniques and topics at the frontiers of current research, and integrates modern concepts and examples from high-energy, statistical, and condensed-matter physics alike. Extensive problem sets further illustrate applications across a range of subfields. The book will be suitable for students across physical subdisciplines who have mastered graduate-level quantum mechanics, and will be a useful reference for researchers"-- Provided by publisher.
Subjects	Quantum field theory.
Notes	Includes bibliographical references and index.
	Description based on print version record and CIP data provided by publisher.
Additional formats	Print version: Fradkin, Eduardo. Quantum field theory Princeton: Princeton University Press, [2021] 9780691149080 (DLC) 2020044826

Quantum field theory: an integrated approach	
LCCN	2020044826
Type of material	Book
Personal name	Fradkin, Eduardo, author.
Main title	Quantum field theory: an integrated approach / Eduardo Fradkin.
Published/Produced	Princeton: Princeton University Press, [2021]
ISBN	9780691149080 (hardback)

	(ebook)
LC classification	QC174.45 .F695 2021
Summary	"Quantum field theory is the mathematical and conceptual framework that describes the physics of the very small, including subatomic particles and quasiparticles. It is used to address a range of problems across subfields, from high-energy physics and gravitation to statistical physics and condensed matter physics. Despite the breadth of its applications, however, the teaching of quantum field theory has historically been strongly oriented toward high-energy physics students, while others-particularly in condensed matter and statistical physics-are typically taught in a separate course, or take an alternate sequence in many-body and statistical physics. Author Eduardo Fradkin strongly believes that this separation is both artificial and detrimental to all groups' understanding of quantum field theory. This textbook, developed from a graduate course Fradkin has taught for decades at the University of Illinois, offers a new, "multicultural" approach to the subject that seeks to remedy this fragmentation. It covers both basic techniques and topics at the frontiers of current research, and integrates modern concepts and examples from high-energy, statistical, and condensed-matter physics alike. Extensive problem sets further illustrate applications across a range of subfields. The book will be suitable for students across physical subdisciplines who have mastered graduate-level quantum mechanics, and will be a useful reference for researchers"-- Provided by publisher.

Subjects	Quantum field theory.
Notes	Includes bibliographical references and index.
Additional formats	Online version: Fradkin, Eduardo Hector, 1950- Quantum field theory Princeton: Princeton University Press, 2021. 9780691189550 (DLC) 2020044827

Quantum field theory: an introduction for chemical physicists	
LCCN	2021017845
Type of material	Book
Personal name	Porter, Richard N. (Richard Needham), 1932- author.
Main title	Quantum field theory: an introduction for chemical physicists / Richard N. Porter, Stony Brook University, USA.
Published/Produced	New Jersey: World Scientific, [2021]
ISBN	9789811239885 (hardcover)
	(ebook)
	(ebook other)
LC classification	QC174.45 .P68 2021
Summary	"This book sets itself apart from existing texts on Quantum Field Theory (QFT) by focusing on chemical applications of QFT such as the modeling of gases and phase transitions, thus catering to an untapped pool of potential users, namely, graduate/upper undergraduate students in chemical physics, as well as chemists and spectroscopists who want an introduction to QFT"-- Provided by publisher.
Subjects	Quantum field theory.
Notes	Includes bibliographical references and index.

Quantum field theory and critical phenomena 5e	
LCCN	2021931817
Type of material	Book
Personal name	Zinn-Justin, Jean, author.
Main title	Quantum field theory and critical phenomena 5e / Jean Zinn-Justin.
Edition	5.
Published/Produced	New York: Oxford University Press, 2021.
ISBN	9780198834625 (hardback)
Summary	"Introduced as a quantum extension of Maxwell's classical theory, quantum electrodynamics has been the first example of a quantum field theory (QFT). Eventually, QFT has become the framework for the discussion of all fundamental interactions at the microscopic scale except, possibly, gravity. More surprisingly, it has also provided a framework for the understanding of second order phase transitions in statistical mechanics. In fact, as hopefully this work illustrates, QFT is the natural framework for the discussion of most systems involving an infinite number of degrees of freedom with local couplings. These systems range from cold Bose gases at the condensation temperature (about ten nanokelvin) to conventional phase transitions (from a few degrees to several hundred) and high energy particle physics up to a TeV, altogether more than twenty orders of magnitude in the energy scale. Therefore, although excellent textbooks about QFT had already been published, I thought, many years ago, that it might not be completely worthless to present a work in which the strong formal relations between particle

	physics and the theory of critical phenomena are systematically emphasized. This option explains some of the choices made in the presentation. A formulation in terms of field integrals has been adopted to study the properties of QFT. The language of partition and correlation functions has been used throughout, even in applications of QFT to particle physics. Renormalization and renormalization group properties are systematically discussed. The notion of effective field theory and the emergence of renormalisable theories are described. The consequences for fine tuning and triviality issue are emphasized. This fifth edition has been updated and fully revised"-- Provided by publisher.
Series	International monographs on physics

Quantum field theory and manifold invariants	
LCCN	2021021514
Type of material	Book
Main title	Quantum field theory and manifold invariants / Daniel S. Freed, Sergei Gukov, Ciprian Manolescu, Constantin Teleman, Ulrike Tillmann, editors.
Published/Produced	Providence: American Mathematical Society, 2021.
ISBN	9781470461232 (hardcover)
	(ebook)
LC classification	QC174.45 .Q3725 2021
Related names	Freed, Daniel S., editor.
	Gukov, Sergei, 1977- editor.
	Manolescu, Ciprian, 1978- editor.
	Teleman, Constantin, 1968- editor.

	Tillmann, Ulrike, 1962- editor.
Subjects	Quantum field theory--Mathematics.
	Invariant manifolds.
	Manifolds and cell complexes -- Differential topology -- Topological quantum field theories (aspects of differential topology).
	Quantum theory -- Quantum field theory; related classical field theories.
Notes	Includes bibliographical references.
Series	IAS/Park City mathematics series, 1079-5634; Volume 28

Restricted congruences in computing	
LCCN	2020019044
Type of material	Book
Personal name	Bibak, Khodakhast, author.
Main title	Restricted congruences in computing / Khodakhast Bibak.
Edition	First edition.
Published/Produced	Boca Raton: CRC Press, 2021.
ISBN	9780367496036 (hardback)
	(ebook)
LC classification	QA608 .B48 2021
Summary	"Congruences are ubiquitous in computer science and related areas. They have interesting applications in signal processing, data structures and algorithms, DNA-based data storage, universal hashing, computational complexity, information theory, coding theory, quantum computing, game theory, discrete mathematics, number theory, cryptography, and more. Therefore, developing techniques for finding (the number of) solutions of congruences is an

	important problem. As the first book of its kind, this book is devoted to studying such problems and their applications. It will be of interest to graduate students and researchers across computer science, electrical engineering, and mathematics"-- Provided by publisher.
Contents	The restricted congruences toolbox -- The GCD-restricted linear congruences -- Applications in universal hashing and authentication with secrecy -- Applications in string theory and quantum field theory -- Alldiff congruences, graph theoretic method, and beyond -- Alldiff congruences meet VT codes -- Binary linear congruence code -- Applications in parallel computing, AI, etc -- Quadratic congruences, Ramanujan graphs, and the Golomb-Welch conjecture.
Subjects	Congruences (Geometry)
	Computer science--Mathematics.
Notes	Includes bibliographical references and index.
Additional formats	Online version: Bibak, Khodakhast. Restricted congruences in computing. First edition Boca Raton: CRC Press, 2021 9781003047179 (DLC) 2020019045

Restricted congruences in computing	
LCCN	2020019045
Type of material	Book
Personal name	Bibak, Khodakhast, author.
Main title	Restricted congruences in computing / Khodakhast Bibak.
Edition	First edition.
Published/Produced	Boca Raton: CRC Press, 2021.

Description	1 online resource
ISBN	9781003047179 (ebook)
	(hardback)
LC classification	QA608
Summary	"Congruences are ubiquitous in computer science and related areas. They have interesting applications in signal processing, data structures and algorithms, DNA-based data storage, universal hashing, computational complexity, information theory, coding theory, quantum computing, game theory, discrete mathematics, number theory, cryptography, and more. Therefore, developing techniques for finding (the number of) solutions of congruences is an important problem. As the first book of its kind, this book is devoted to studying such problems and their applications. It will be of interest to graduate students and researchers across computer science, electrical engineering, and mathematics"-- Provided by publisher.
Contents	The restricted congruences toolbox -- The GCD-restricted linear congruences -- Applications in universal hashing and authentication with secrecy -- Applications in string theory and quantum field theory -- Alldiff congruences, graph theoretic method, and beyond -- Alldiff congruences meet VT codes -- Binary linear congruence code -- Applications in parallel computing, AI, etc -- Quadratic congruences, Ramanujan graphs, and the Golomb-Welch conjecture.
Subjects	Congruences (Geometry)
	Computer science--Mathematics.
Notes	Includes bibliographical references and index.

	Description based on print version record and CIP data provided by publisher; resource not viewed.
Additional formats	Print version: Bibak, Khodakhast. Restricted congruences in computing First edition. Boca Raton: CRC Press, 2021. 9780367496036 (DLC) 2020019044
Roman jackiw: 80th birthday festschrift	
LCCN	2020937448
Type of material	Book
Main title	Roman jackiw: 80th birthday festschrift / Antti Niemi, Terry Tomboulis, Kok Khoo Phua.
Published/Produced	Hackensack: World Scientific Publishing Co. Pte. Ltd., 2020.
ISBN	9789811210662 (hardcover)
Related names	Niemi, Antti, editor.
	Tomboulis, Terry, editor.
	Phua, Kok Khoo, editor.
Summary	"Professor Roman Jackiw is a theoretical physicist renowned for his many fundamental contributions and discoveries in quantum and classical field theories, ranging from high energy physics and gravitation to condensed matter and the physics of fluids. Among his major achievements is the establishment of the presence of the famous Adler-Bell-Jackiw anomalies in quantum field theory, a discovery with far-reaching implications for the structure of the Standard Model of particle physics and all attempts to go beyond it. Other important contributions, among many, that one may mention here are the topological mass term in

	gravity and gauge theories, and the fractionalization of fermion number and charge in the presence of topological objects. Roman Jackiw, a Professor Emeritus at the MIT Center for Theoretical Physics, is the recipient of several international awards including the Dannie Heineman Prize for Mathematical Physics and the Dirac Medal of the ICTP. He is a member of the US National Academy of Sciences and honorary doctor of Kiev, Montreal, Tours, Turin and Uppsala universities. To celebrate his 80th birthday, many students and colleagues of Professor Jackiw have come together to share interesting anecdotes of working with him as well as their latest research, some of it inspired by his work. Edited by his former students Antti Niemi and Terry Tomboulis together with his long-time friend KK Phua, this festschrift volume is a must-have collection for all theoretical physicists"-- Provided by publisher.

Scientific realism and the quantum	
LCCN	2019951277
Type of material	Book
Main title	Scientific realism and the quantum / edited by Steven French and Juha Saatsi.
Edition	First edition.
Published/Produced	Oxford: Oxford University Press, 2020.
	©2020
Description	viii, 320 pages: illustrations; 24 cm
ISBN	9780198814979 (hardcover)
	0198814976 (hardcover)
LC classification	QC6 .S4367 2020

Related names	French, Steven, editor.
	Saatsi, Juha, editor.
Summary	"Quantum theory is widely regarded as one of the most successful theories in the history of science. It explains a hugely diverse array of phenomena and is a natural candidate for our best representation of the world at the level of 'fundamental' physics. But how can the world be the way quantum theory says it is? It is famously unclear what the world is like according to quantum physics, which presents a serious problem for the scientific realist who is committed to regarding our best theories as more or less true. The present volume canvasses a variety of responses to this problem, from restricting or revising realism in different ways to exploring entirely new directions in the lively debate surrounding realist interpretations of quantum physics. Some urge us to focus on new formulations of the theory itself, while others examine the status of scientific realism in the further context of quantum field theory. Each chapter is written by a renowned specialist in the field and is aimed at graduate students and researchers in both physics and the philosophy of science. Together they offer a range of illuminating new perspectives on this fundamental debate and exemplify the fruitful interaction between physics and philosophy." -- Provided by publisher.
Contents	Introduction / Steven French and Juha Saatsi -- Part I: Rethinking scientific realism -- Scientific realism without the quantum / Carl Hoefer -- Truth vs. progress realism about spin / Juha

	Saatsi -- Part II: Underdetermination and interpretation -- Can we quarantine the quantum blight? / Craig Callender -- On the plurality of quantum theories: Quantum theory as a framework and its implications for the quantum measurement problem / David Wallace -- Naturalism and the interpretation of quantum mechanics / J. E. Wolff -- Part III: Pragmatism about quantum theory -- Pragmatist quantum theory handle objectivity about explanations? / Lina Jansson -- Quantum mechanics and its (dis)contents / Peter J. Lewis -- Part IV: Wavefunctions and quantum state realism -- Losing sight of the forest for the psi: Beyond the wavefunction hegemony / Alisa Bokulich -- Scientific realism without the wave function / Valia Allori -- On the status of quantum state realism / Wayne C. Myrvold -- Part V: Scientific realism and quantum field theory -- The non-miraculous success of formal analogies in quantum theories / Doreen Fraser -- Toward a realist view of quantum field theory / James D. Fraser -- Perturbing realism / Laura Ruetsche.
Subjects	Physics--Philosophy.
	Physics--Philosophy.
Notes	Includes bibliographical references and indexes.

Semiclassical and stochastic gravity: quantum field effects on curved spacetime	
LCCN	2019038185
Type of material	Book
Personal name	Hu, B. L. (Bei-Lok), author.

Main title	Semiclassical and stochastic gravity: quantum field effects on curved spacetime / Bei-Lok B. Hu and Enric Verdaguer.
Published/Produced	[New York, NY]: [Cambridge University Press], [2020]
Description	1 online resource
ISBN	9780511667497 (epub)
	(hardback)
LC classification	QC178
Related names	Verdaguer, E. (Enric), author.
Summary	"The two pillars of modern physics are general relativity and quantum field theory, the former describes the large scale structure and dynamics of space-time, the latter, the microscopic constituents of matter. Combining the two yields quantum field theory in curved space-time, which is needed to understand quantum field processes in the early universe and black holes, such as the well-known Hawking effect. This book examines the effects of quantum field processes back-reacting on the background space-time which become important near the Planck time (10-43 sec). It explores the self-consistent description of both space-time and matter via the semiclassical Einstein equation of semiclassical gravity theory, exemplified by the inflationary cosmology, and fluctuations of quantum fields which underpin stochastic gravity, necessary for the description of metric fluctuations (space-time foams). Covering over four decades of thematic development, this book is a valuable resource for researchers interested in quantum field theory, gravitation and cosmology"-- Provided by publisher.

Contents	Overview: Main themes. Key issues. Reader's guide -- 'In-out' effective action. Dimensional regularization -- 'In-in' effective action. Stress tensor. Thermal fields -- Stress-energy tensor and correlators: zeta-function method -- Stress-energy tensor and correlation: point separation -- Infrared behavior of interacting quantum fields -- Advanced field theory topics -- Backreaction of early universe quantum processes -- Metric correlations at one-loop: in-in and large N -- The Einstein-Langevin equation -- Metric fluctuations in Minkowski spacetime -- Cosmological backreaction with fluctuations -- Structure formation in the early universe -- Black hole backreaction and fluctuations -- Stress-energy tensor fluctuations in de Sitter space -- Two-point metric perturbations in de Sitter -- Riemann tensor correlator in de Sitter -- Epilogue: Linkage with quantum gravity.
Subjects	Quantum gravity.
	Space and time.
Notes	Includes bibliographical references and index.
	Description based on print version record and CIP data provided by publisher; resource not viewed.
Additional formats	Print version: Hu, B. L. (Bei-Lok). Semiclassical and stochastic gravity [New York, NY]: [Cambridge University Press], [2020] 9780521193573 (DLC) 2019038184

Semiclassical and stochastic gravity: quantum field effects on curved spacetime	
LCCN	2019038184
Type of material	Book
Personal name	Hu, B. L. (Bei-Lok), author.
Main title	Semiclassical and stochastic gravity: quantum field effects on curved spacetime / Bei-Lok B. Hu and Enric Verdaguer.
Published/Produced	[New York, NY]: [Cambridge University Press], [2020]
ISBN	9780521193573 (hardback)
	(epub)
LC classification	QC178 .H822 2020
Related names	Verdaguer, E. (Enric), author.
Summary	"The two pillars of modern physics are general relativity and quantum field theory, the former describes the large scale structure and dynamics of space-time, the latter, the microscopic constituents of matter. Combining the two yields quantum field theory in curved space-time, which is needed to understand quantum field processes in the early universe and black holes, such as the well-known Hawking effect. This book examines the effects of quantum field processes back-reacting on the background space-time which become important near the Planck time (10-43 sec). It explores the self-consistent description of both space-time and matter via the semiclassical Einstein equation of semiclassical gravity theory, exemplified by the inflationary cosmology, and fluctuations of quantum fields which underpin stochastic gravity, necessary for the description of metric

	fluctuations (space-time foams). Covering over four decades of thematic development, this book is a valuable resource for researchers interested in quantum field theory, gravitation and cosmology"-- Provided by publisher.
Contents	Overview: Main themes. Key issues. Reader's guide -- 'In-out' effective action. Dimensional regularization -- 'In-in' effective action. Stress tensor. Thermal fields -- Stress-energy tensor and correlators: zeta-function method -- Stress-energy tensor and correlation: point separation -- Infrared behavior of interacting quantum fields -- Advanced field theory topics -- Backreaction of early universe quantum processes -- Metric correlations at one-loop: in-in and large N -- The Einstein-Langevin equation -- Metric fluctuations in Minkowski spacetime -- Cosmological backreaction with fluctuations -- Structure formation in the early universe -- Black hole backreaction and fluctuations -- Stress-energy tensor fluctuations in de Sitter space -- Two-point metric perturbations in de Sitter -- Riemann tensor correlator in de Sitter -- Epilogue: Linkage with quantum gravity.
Subjects	Quantum gravity.
	Space and time.
Notes	Includes bibliographical references and index.
Additional formats	Online version: Hu, B. L. (Bei-Lok) Semiclassical and stochastic gravity. [Cambridge]; [New York, NY]: [Cambridge University Press], [2020] 9780511667497 (DLC) 2019038185

Skyrmions: a theory of nuclei	
LCCN	2021062774
Type of material	Book
Personal name	Manton, Nicholas, 1952- author.
Main title	Skyrmions: a theory of nuclei / Nicholas S. Manton.
Published/Produced	Hackensack, New Jersey: World Scientific, [2022]
ISBN	9781800612471 (hardcover)
	(ebook for institutions)
	(ebook for individuals)
LC classification	QC793.3.S8 M36 2022
Summary	"Skyrmions - A Theory of Nuclei surveys 60 years of research into the brilliant and imaginative idea of Tony Skyrme that atomic nuclei can be modelled as topologically stable states, known as Skyrmions, in an effective quantum field theory of pions. Skyrme theory emerges as a low-energy approximation to the more fundamental theory of quarks and gluons - quantum chromodynamics (QCD). Skyrmions give spatial structure to the protons and neutrons inside nuclei, and capture the interactions of these basic particles, allowing them to partially merge. Skyrme theory also gives a topological explanation for the conservation of baryon number, a fundamental principle of physics. The book summarises the particle and field theory background, then presents Skyrme field theory together with the mathematics needed to understand it. Many beautiful and surprisingly symmetric Skyrmions are described and illustrated in colour. Quantized Skyrmion

	motion models the momentum, energy and spin of nuclei, and also their isospin, the quantum number distinguishing protons and neutrons. Skyrmion vibrations also need to be quantized, and the book reviews how the complicated energy spectra of several nuclei, including Carbon-12 and Oxygen-16, are accurately modelled by rotational/vibrational states of Skyrmions. A later chapter explores variants of Skyrme theory, incorporating mesons heavier than pions, and extending the basic theory to include particles like kaons that contain strange quarks. The final chapter introduces the Sakai-Sugimoto model, which relates Skyrmions to gauge theory instantons in a higher-dimensional framework inspired by string theory"-- Provided by publisher.
Contents	Foreword / by Prof. K. K. Phua -- Fields and particles -- Lagrangians and symmetries -- Skyrme theory -- Quantization of Skyrmions -- Skyrmions with higher B: massless pions -- Rigid-Body Skyrmion quantization -- Skyrmions with higher B: massive pions -- Quantized Skyrmions with even B [less than or equal to] 12 -- Modelling oxygen-16 -- Modelling calcium-40 -- Electromagnetic transition strengths -- The Sakai-Sugimoto model.
Subjects	Skyrme model.
	Atomic structure.
	Topology.
	Nuclear physics--History.
	Quantum theory--History.
Notes	Includes bibliographical references and index.

Standard model phenomenology	
LCCN	2021059303
Type of material	Book
Personal name	Khalil, Shaaban, author.
Main title	Standard model phenomenology / Shaaban Khalil & Stefano Moretti.
Edition	First edition.
Published/Produced	Boca Raton: CRC Press, 2022.
ISBN	9781138336438 (hardback)
	9781138336438 (paperback)
	(ebook)
LC classification	QC794.6.S75 K43 2022
Related names	Moretti, Stefano, 1966- author.
Summary	"This new book is fully up to date with all the latest developments on both theoretical and experimental investigations of the Standard Model (SM) of particle physics with a particular emphasis on its historical development on both sides. It further stresses the cross-fertilisation between the two sub-disciplines of theoretical and experimental particle physics which has been instrumental in establishing the SM. In other words, the book develops a truly phenomenological attitude to the subject. In addition to emphasising the successes of the SM, this book also critically assesses its limitations and raises key unanswered questions for the purpose of presenting a new perspective of how to further our knowledge above and beyond it. It also contains both historical information from past experiments and latest results from the Large Hadron Collider at CERN. This book will be an invaluable reference to advanced

	undergraduate and postgraduate students, in addition to early-stage researchers in the field. Key Features: Provides a unique approach not found in current literature in developing and verifying the SM Presents the theory pedagogically but rigorously from basic knowledge of quantum field theory Brings together experimental and theoretical practice in one, cohesive text"-- Provided by publisher.
Subjects	Standard model (Nuclear physics)
	Particles (Nuclear physics)
Notes	Includes bibliographical references and index.

Structural aspects of quantum field theory and noncommutative geometry	
LCCN	2021013061
Type of material	Book
Personal name	Grensing, Gerhard, author.
Main title	Structural aspects of quantum field theory and noncommutative geometry / Gerhard Grensing. University of Kiel, Germany.
Edition	Second edition.
Published/Produced	New Jersey: World Scientific, 2021.
ISBN	9789811237010 (set)
	9789811238000 (v. 1; hardcover)
	9789811238017 (v. 2; hardcover)
LC classification	QC174.45 .G724 2021
Summary	"The book is devoted to the subject of quantum field theory. It is divided into two volumes. The first volume can serve as a textbook on main techniques and results of quantum field theory, while the second treats more recent developments, in particular the subject of

	quantum groups and noncommutative geometry, and their interrelation. The second edition is extended by additional material, mostly concerning the impact of noncommutative geometry on theories beyond the standard model of particle physics, especially the possible role of torsion in the context of the dark matter problem. Furthermore, the text includes a discussion of the Randall-Sundrum model and the Seiberg-Witten equations"-- Provided by publisher.
Contents	v. 1. -- v. 2.
Subjects	Quantum field theory.

Student friendly quantum field theory the standard model	
LCCN	2021949566
Type of material	Book
Personal name	Klauber, Robert Douglas, 1943- author.
Main title	Student friendly quantum field theory the standard model / Robert D. Klauber.
Edition	First.
Published/Produced	Fairfield: Sandtrove Press, 2021.
ISBN	9780984513970 (v. 2; hardback)
	9780984513987 (v. 2; paperback)
Summary	"A pedagogic introduction to the standard model of elementary particles and fields. Volume 1 presented basic principles of quantum field theory and applied those principles to quantum electrodynamics. This Volume 2 applies those principles to electroweak and strong interactions"-- Provided by publisher.

The Einstein-Klein-Gordon coupled system: global stability of the Minkowski solution	
LCCN	2021050082
Type of material	Book
Personal name	Ionescu, Alexandru Dan, 1973- author.
Main title	The Einstein-Klein-Gordon coupled system: global stability of the Minkowski solution / Alexandru D. Ionescu and Benoît Pausader.
Published/Produced	Princeton: Princeton University Press, [2022]
Description	1 online resource
ISBN	9780691233031 (ebook)
	(hardback)
	(paperback)
LC classification	QC174.26.W28
Related names	Pausader, Benoît, 1982- author.
Summary	"This monograph presents a significant new result in general relativity. In particular, it provides a proof related to the Einstein-Klein-Gordon equation, a fundamental equation in mathematical physics that couples the Einstein equation of general relativity with a matter field described by the Klein-Gordon equation. The book begins with an introduction and history of the subject, proceeds to prove several auxiliary lemmas, and culminates in the central proof. This book represents a significant advance in mathematical physics, and provides the most cutting-edge treatment of the Einstein-Klein-Gordon equation to date"-- Provided by publisher.
Contents	The main construction and outline of the proof -- Preliminary estimates -- The nonlinearities $N^h / [\infty][\beta]$ and $n^{[\psi]}$ -- Improved energy

	estimates -- Improved profile bounds -- The main theorems.
Subjects	Klein-Gordon equation.
	General relativity (Physics)
	Quantum field theory.
	Mathematical physics.
	Science / Physics / Mathematical & Computational
	Mathematics / General
Notes	Includes bibliographical references and index.
	Description based on print version record and CIP data provided by publisher; resource not viewed.
Additional formats	Print version: Ionescu, Alexandru Dan, 1973- Einstein-Klein-Gordon coupled system Princeton: Princeton University Press, [2022] 9780691233055 (DLC) 2021050081
Series	Annals of mathematics studies; number 213

The Einstein-Klein-Gordon coupled system: global stability of the Minkowski solution	
LCCN	2021050081
Type of material	Book
Personal name	Ionescu, Alexandru Dan, 1973- author.
Main title	The Einstein-Klein-Gordon coupled system: global stability of the Minkowski solution / Alexandru D. Ionescu and Benoît Pausader.
Published/Produced	Princeton: Princeton University Press, [2022]
ISBN	9780691233055 (hardback)
	9780691233048 (paperback)
	(ebook)
LC classification	QC174.26.W28 I583 2022
Related names	Pausader, Benoît, 1982- author.

Summary	"This monograph presents a significant new result in general relativity. In particular, it provides a proof related to the Einstein-Klein-Gordon equation, a fundamental equation in mathematical physics that couples the Einstein equation of general relativity with a matter field described by the Klein-Gordon equation. The book begins with an introduction and history of the subject, proceeds to prove several auxiliary lemmas, and culminates in the central proof. This book represents a significant advance in mathematical physics, and provides the most cutting-edge treatment of the Einstein-Klein-Gordon equation to date"-- Provided by publisher.
Contents	The main construction and outline of the proof -- Preliminary estimates -- The nonlinearities $N^h / [\infty][\beta]$ and $n^{[\psi]}$ -- Improved energy estimates -- Improved profile bounds -- The main theorems.
Subjects	Klein-Gordon equation.
	General relativity (Physics)
	Quantum field theory.
	Mathematical physics.
	Science / Physics / Mathematical & Computational
	Mathematics / General
Notes	Includes bibliographical references and index.
Additional formats	Online version: Ionescu, Alexandru Dan., 1973- Einstein-Klein-Gordon coupled system Princeton: Princeton University Press, [2022] 9780691233031 (DLC) 2021050082
Series	Annals of mathematics studies; number 213

The world in the wave function: a metaphysics for quantum physics	
LCCN	2020024392
Type of material	Book
Personal name	Ney, Alyssa, author.
Main title	The world in the wave function: a metaphysics for quantum physics / Alyssa Ney.
Published/Produced	New York, NY: Oxford University Press, [2020]
ISBN	9780190097714 (hardback)
	(epub)
LC classification	QC174.26.W3 N49 2020
Summary	"What are the ontological implications of quantum theories, that is, what do they tell us about the fundamental objects that make up our world? How should quantum theories make us reevaluate our classical conceptions of the basic constitution of material objects and ourselves? Is there fundamental quantum nonlocality? This book articulates several rival approaches to answering these questions, ultimately defending the wave function realist approach. It is a way of interpreting quantum theories so that the central object they describe is the quantum wave function, interpreted as a field, and that the nonseparability and nonlocality we seem to find in quantum mechanics are ultimately manifestations of a more intuitive, separable and local picture in higher dimensions. quantum mechanics, quantum field theory, wave function, wave function realism, measurement problem, macro-object problem, primitive ontology, quantum entanglement, quantum nonlocality, quantum ontology"-- Provided by publisher.
Subjects	Wave functions.

	Quantum theory.
Notes	Includes bibliographical references and index.
Additional formats	Online version: Ney, Alyssa, The world in the wave function New York: Oxford University Press, 2020. 9780190097738 (DLC) 2020024393

The world in the wave function: a metaphysics for quantum physics	
LCCN	2020024393
Type of material	Book
Personal name	Ney, Alyssa, author.
Main title	The world in the wave function: a metaphysics for quantum physics / Alyssa Ney.
Published/Produced	New York, NY: Oxford University Press, [2020]
Description	1 online resource
ISBN	9780190097738 (epub)
	(hardback)
LC classification	QC174.26.W3
Summary	"What are the ontological implications of quantum theories, that is, what do they tell us about the fundamental objects that make up our world? How should quantum theories make us reevaluate our classical conceptions of the basic constitution of material objects and ourselves? Is there fundamental quantum nonlocality? This book articulates several rival approaches to answering these questions, ultimately defending the wave function realist approach. It is a way of interpreting quantum theories so that the central object they describe is the quantum wave function, interpreted as a field, and that the nonseparability and nonlocality we seem to find in quantum mechanics are ultimately

	manifestations of a more intuitive, separable and local picture in higher dimensions. quantum mechanics, quantum field theory, wave function, wave function realism, measurement problem, macro-object problem, primitive ontology, quantum entanglement, quantum nonlocality, quantum ontology"-- Provided by publisher.
Subjects	Wave functions.
	Quantum theory.
Notes	Includes bibliographical references and index.
	Description based on print version record and CIP data provided by publisher; resource not viewed.
Additional formats	Print version: Ney, Alyssa. The world in the wave function New York, NY: Oxford University Press, [2020] 9780190097714 (DLC) 2020024392

Topological phases of matter and quantum computation: AMS Special Session on Topological Phases of Matter and Quantum Computation, September 24-25, 2016, Bowdoin College, Brunswick, Maine	
LCCN	2019040079
Type of material	Book
Meeting name	AMS Special Session on Topological Phases of Matter and Quantum Computation (2016: Brunswick, Me.), author.
Main title	Topological phases of matter and quantum computation: AMS Special Session on Topological Phases of Matter and Quantum Computation, September 24-25, 2016, Bowdoin College, Brunswick, Maine / Paul Bruillard, Carlos Ortiz Marrero, Julia Plavnik, editors.

Published/Produced	Providence, Rhode Island: American Mathematical Society, [2020]
Description	xii, 226 pages: illustrations; 26 cm
ISBN	9781470440749 (paperback)
	(ebook)
LC classification	QA76.889 .A467 2020
Related names	Bruillard, Paul, 1984- editor.
	Ortiz Marrero, Carlos, 1989- editor.
	Plavnik, Julia, 1985- editor.
Contents	Lie theory for fusion categories: a research primer / Andrew Schopieray -- Entanglement and the Temperley-Lieb category / Michael Brannan and Benoit Collins -- Lifting shadings on symmetrically self-dual subfactor planar algebras / Zhengwei Liu, Scott Morrison, and David Penneys -- Q-systems and compact W*-algebra objects / Corey Jones and David Penneys -- Dimension as a quantum statistic and the classification of metaplectic categories / Paul Bruillard, Paul Gustafson, Julia Plavnik, and Eric Rowell -- The rank of G-crossed braided extensions of modular tensor categories / Marcel Bischoff -- Symmetry defects and their application to topological quantum computing / Colleen Delaney and Zhenghan Wang -- Topological quantum computation with gapped boundaries and boundary defects / Iris Cong and Zhenghan Wang -- Classification of gapped quantum liquid phases of matter / Xiao-Gang Wen -- Schur-type invariants of branched G-covers of surfaces / Eric Samperton -- Quantum error-correcting codes over finite Frobenius rings / Andreas Klappenecker, Sangjun Lee, and Andrew

	Nemec -- A short history of frames and quantum designs / Bernhard Bodmann and John Haas.
Subjects	Quantum computing--Congresses.
	Topological groups--Congresses.
	Quantum groups--Congresses.
	Categories (Mathematics)--Congresses.
	Quantum theory -- Groups and algebras in quantum theory -- Quantum groups and related algebraic methods.
	Associative rings and algebras -- Modules, bimodules and ideals -- Module categories; module theory in a category-theoretic context; Morit
	Quantum theory -- Quantum field theory; related classical field theories -- Axiomatic quantum field theory; operator algebras.
	Group theory and generalizations -- Linear algebraic groups and related topics.
	Category theory; homological algebra
	K-theory -- Higher algebraic K-theory -- Symmetric monoidal categories.
Notes	Includes bibliographical references.
Series	Contemporary mathematics, 0271-4132; volume 747

Topological phases of matter and quantum computation: AMS Special Session on Topological Phases of Matter and Quantum Computation, September 24-25, 2016, Brunswick, Maine	
LCCN	2019723503
Type of material	Book
Meeting name	AMS Special Session on Topological Phases of Matter and Quantum Computation (2016: Brunswick, Me.), author.

Main title	Topological phases of matter and quantum computation: AMS Special Session on Topological Phases of Matter and Quantum Computation, September 24-25, 2016, Brunswick, Maine / Paul Bruillard, Carlos Ortiz Marrero, Julia Plavnik, editors.
Published/Produced	Providence, Rhode Island: American Mathematical Society, [2020]
Description	1 online resource
ISBN	9781470454579 (ebook)
	(paperback)
LC classification	QA76.889
Related names	Bruillard, Paul, 1984- editor.
	Ortiz Marrero, Carlos, 1989- editor.
	Plavnik, Julia, 1985- editor.
Contents	Lie theory for fusion categories: a research primer / Andrew Schopieray -- Entanglement and the Temperley-Lieb category / Michael Brannan and Benoit Collins -- Lifting shadings on symmetrically self-dual subfactor planar algebras / Zhengwei Liu, Scott Morrison, and David Penneys -- Q-systems and compact W*-algebra objects / Corey Jones and David Penneys -- Dimension as a quantum statistic and the classification of metaplectic categories / Paul Bruillard, Paul Gustafson, Julia Plavnik, and Eric Rowell -- The rank of G-crossed braided extensions of modular tensor categories / Marcel Bischoff -- Symmetry defects and their application to topological quantum computing / Colleen Delaney and Zhenghan Wang -- Topological quantum computation with gapped boundaries and boundary defects / Iris Cong and

	Zhenghan Wang -- Classification of gapped quantum liquid phases of matter / Xiao-Gang Wen -- Schur-type invariants of branched G-covers of surfaces / Eric Samperton -- Quantum error-correcting codes over finite Frobenius rings / Andreas Klappenecker, Sangjun Lee, and Andrew Nemec -- A short history of frames and quantum designs / Bernhard Bodmann and John Haas.
Subjects	Quantum computing--Congresses.
	Topological groups--Congresses.
	Quantum groups--Congresses.
	Categories (Mathematics)--Congresses.
	Quantum theory -- Groups and algebras in quantum theory -- Quantum groups and related algebraic methods.
	Associative rings and algebras -- Modules, bimodules and ideals -- Module categories; module theory in a category-theoretic context; Morit
	Quantum theory -- Quantum field theory; related classical field theories -- Axiomatic quantum field theory; operator algebras.
	Group theory and generalizations -- Linear algebraic groups and related topics.
	Category theory; homological algebra
	\$K\$-theory -- Higher algebraic \$K\$-theory -- Symmetric monoidal categories.
Notes	Includes bibliographical references.
	Description based on print version record.
Additional formats	Print version: Topological phases of matter and quantum computation Providence, Rhode Island: American Mathematical Society, [2020] 9781470440749 (DLC) 2019040079

Series	Contemporary mathematics, 0271-4132; volume 747

War or peaceful transformation: multidisciplinary and international perspectives	
LCCN	2019048606
Type of material	Book
Main title	War or peaceful transformation: multidisciplinary and international perspectives / Marek J. Celinski, Kathryn M. Gow, editors.
Published/Produced	Hauppauge, NY: Nova Science Publishers, 2020.
Description	1 online resource
ISBN	9781536165951 (adobe pdf)
	(hardcover)
LC classification	JZ5538
Related names	Celinski, Marek J., editor.
	Gow, Kathryn, editor.
Summary	"Concern for humanity's future has never been more urgent than now - in the present time - when humanity has achieved the level of capability of destroying itself either through environmental disasters or nuclear wars. On the other hand, we have also achieved material and psychological knowledge and progress that can assist us in understanding not only the causation, but also the potential embedded in human nature, to choose either the path to self-destruction or to sustained peace. In this book, we present both the ubiquitous causes of violent discontent and wars and successful attempts to reduce or resolve conflict. Our authors from five continents represent historic, military, philosophical, socio-

	political, and psychological perspectives and address some of the important issues which any peace-oriented initiative or society at large must contend with. These refer to access to natural resources, ethnicity, religion, human rights, political systems (whether democratic or autocratic), differences in political and military strength and WMDs, and aspirations of the leaders - in combination with the ubiquitous need for control through domination, historic traditions (such as glorification of war effort as heroism and as a sacrifice in the name of lofty ideas). We offer a vision of a humanistic approach to promote peaceful problem solving that needs to be propagated by education, media, political programs and diplomacy in order to lead to peaceful transformations. The role of the military is given special attention. The novelty of our approach is that we address the typical life situations leading to social unrest and wars within the context of the human mind's capabilities to deal with life challenges. Our Challenge-Resilience-Resourcefulness-Wisdom model (previously published by NOVA) shows how we can study and analyse human errors, regressive tendencies and limitations in order to reframe them as an inspiration for optimal and wise decisions. Based on our authors' insights, we provide many descriptions of how to deal with social adversity in different locations of the world and also examples of actual successes and failures of peaceful transformations. These chapters provide important knowledge and tools

	for a wide range of professionals"-- Provided by publisher.
Contents	Human nature and its potential for war and peace / Marek J. Celinski, Andrzej R. Celinski and Kathryn M. Gow -- War, peace, and conflict resolution in the classical world / Jaime A. González-Ocaña -- Ethnic identity, resources, control and supremacy: a brief history of early South African conflicts / Jacques J. Gouws -- Jihad: peaceful definitions and applications / Norman C. Rothman -- Fear and loathing: tribalism in the age of the Internet / Dipak K. Gupta -- The war on drugs: a struggle for the human soul / Andrzej R. Celinski -- Eritrea: a failed state and victim of sellout diplomacy / Tseggai Isaac -- Shattered hopes: the disintegration of South Africa's peaceful transition / Jacques J. Gouws -- Colombia in trauma: a conflict and post-conflict scenario / Saúl M. Rodríguez and Fabio Sánchez -- The Central European experience of war and peace: the nonviolent Czech case / Martina Klicperova-Baker -- The European Union: a case study in peace / Stephen T. Satkiewicz -- Mutual assured destruction as a strategy for peace / João José Brandão Ferreira -- Conflict resolution and peace building: cultural barriers and facilitators / Harsheeta Razora -- Resolution of international and civil war conflicts by diplomatic and military means / Harkirat Singh -- Ideological and policy alternatives to the resolution of Africa's perpetual crisis: is there a worthy policy or ideological alternative? / Tseggai Isaac -- Coping with violence and adversity: general typology

	and concrete illustrations on Czech case / Martina Klicperova-Baker -- Orientations toward achievable world peace / Frank J. Lucatelli and Nancy Ann Hayes -- Building peace in times of conflict: examining military psychology through Gandhi's lens / Swati Mukherjee -- Consciousness: the bridge between war and peace / Sandeep Gupta and Anand Shankar -- Polemology: the pursuit for lasting peace / Jacques J. Gouws -- Virtue as a basis for non-violence and creative maladjustment: humanistic and positive psychological solutions to war and violence / Brent Dean Robbins -- An integrated quantum field theory of cosmos, consciousness and algorithmic intelligence to promote peace / Ernest Lawrence Rossi and Kathryn Lane Rossi -- Education for peace and conflict resolution / Diana Vladimirovna Prokofyeva -- Is peace achievable? / Marek J. Celinski, Andrzej R. Celinski and Kathryn M. Gow.
Subjects	Peacebuilding--Case studies.
	Peaceful change (International relations)--Case studies.
	Conflict resolution--Case studies.
Notes	Includes bibliographical references and index.
	Description based on print version record and CIP data provided by publisher; resource not viewed.
Additional formats	Print version: War or peaceful transformation Hauppauge, NY: Nova Science Publishers, 2020. 9781536165944 (DLC) 2019048605
Series	Terrorism, hot spots and conflict-related issues

What is a quantum field theory?: a first introduction for mathematicians	
LCCN	2021020787
Type of material	Book
Personal name	Talagrand, Michel, 1952- author.
Main title	What is a quantum field theory?: a first introduction for mathematicians / Michel Talagrand.
Edition	First edition.
Published/Produced	New York: Cambridge University Press, 2021.
Description	1 online resource
ISBN	9781108225144 (epub)
	(hardback)
LC classification	QC174.45
Summary	"Quantum field theory (QFT) is one of the great achievements of physics, of profound interest to mathematicians. Most pedagogical texts on QFT are geared toward budding professional physicists, however, whereas mathematical accounts are abstract and difficult to relate to the physics. This book bridges the gap. While the treatment is rigorous whenever possible, the accent is not on formality but on explaining what the physicists do and why, using precise mathematical language. In particular, it covers in detail the mysterious procedure of renormalization. Written for readers with a mathematical background but no previous knowledge of physics and largely self-contained, it presents both basic physical ideas from special relativity and quantum mechanics and advanced mathematical concepts in complete detail. It will be of interest to mathematicians wanting to learn

	about QFT and, with nearly 300 exercises, also to physics students seeking greater rigor than they typically find in their courses"-- Provided by publisher.
Subjects	Quantum field theory.
	Science / Physics / Mathematical & Computational
Notes	Includes bibliographical references and index.
	Description based on print version record and CIP data provided by publisher; resource not viewed.
Additional formats	Print version: Talagrand, Michel, 1952- What is a quantum field theory? First edition. New York: Cambridge University Press, 2021 9781316510278 (DLC) 2021020786

What is a quantum field theory?: a first introduction for mathematicians	
LCCN	2021020786
Type of material	Book
Personal name	Talagrand, Michel, 1952- author.
Main title	What is a quantum field theory?: a first introduction for mathematicians / Michel Talagrand.
Edition	First edition.
Published/Produced	New York: Cambridge University Press, 2021.
ISBN	9781316510278 (hardback)
	(epub)
LC classification	QC174.45 .T35 2021
Summary	"Quantum field theory (QFT) is one of the great achievements of physics, of profound interest to mathematicians. Most pedagogical texts on QFT are geared toward budding professional

	physicists, however, whereas mathematical accounts are abstract and difficult to relate to the physics. This book bridges the gap. While the treatment is rigorous whenever possible, the accent is not on formality but on explaining what the physicists do and why, using precise mathematical language. In particular, it covers in detail the mysterious procedure of renormalization. Written for readers with a mathematical background but no previous knowledge of physics and largely self-contained, it presents both basic physical ideas from special relativity and quantum mechanics and advanced mathematical concepts in complete detail. It will be of interest to mathematicians wanting to learn about QFT and, with nearly 300 exercises, also to physics students seeking greater rigor than they typically find in their courses"-- Provided by publisher.
Subjects	Quantum field theory.
	Science / Physics / Mathematical & Computational
Notes	Includes bibliographical references and index.
Additional formats	Online version: Talagrand, Michel, 1952- What is a quantum field theory? First edition. New York: Cambridge University Press, 2021 9781108225144 (DLC) 2021020787

Wilson lines in quantum field theory	
LCCN	2019951784
Type of material	Book
Personal name	Cherednikov, Igor Olegovich, author.

Main title	Wilson lines in quantum field theory / Igor O. Cherednikov, Tom Mertens, Frederik Van der Veken.
Edition	2nd edition.
Published/Produced	Berlin; Boston: De Gruyter, [2020]
Description	XI, 274 pages: illustrations; 25 cm.
ISBN	9783110650921 (hardback)
	(pdf)
	(epub)
LC classification	QA174.2 .C45 2020
Related names	Mertens, Tom, author.
	Veken, Frederik F. Van der, author.
Subjects	Loops (Group theory)
	Quantum field theory--Mathematics.
	Gauge fields (Physics)
Notes	Includes bibliographical references (pages 266-268) and index.
Series	De gruyter studies in mathematical physics, 2194-3532; Volume 24

INDEX

A

B

C

Q

R

S

T

Y